NOVEMBER

FALL CONVOCATION

Southern Alberta Jubilee Auditorium

1961

THE PRESENTATION

The Dean of the Faculty of Arts and Science asks the candidates for degrees to rise, and presents them with these words:

"Eminent Chancellor and President, on behalf of the Faculties, I present to you these scholars and ask that they be pledged and admitted to the degrees to which they are entitled."

THE PLEDGE

The President, then rising, addresses the candidates in these words:

"Do you promise faithfully to observe and loyally to maintain the statutes, customs, privileges and liberties of this your University? Do you pledge your honor that when you become a member of Convocation you will vote only for those whom you surely know or firmly believe to be fit and proper persons to share in the government of this University? Do you solemnly promise to conduct yourself in all things loyally and faithfully to the honor of your University, the encouragement of learning and the good of your country?"

Each candidate replies: "These things I pledge myself to do."

THE ADMISSION

The Chancellor then addresses the candidates in these words:

"By virtue of the authority vested in me by the Legislature of this Province, and with the consent of this University, I admit you to the degrees to which you are entitled, and invest you with all the powers, rights and privileges pertaining to such degrees, and I charge you to use them for the glory of God and honor of your country."

Candidates then sit down.

University of Alberta

FALL CONVOCATION

for the

CONFERRING OF DEGREES

Two-fifteen o'clock, Saturday afternoon
November the Eighteenth
Nineteen Hundred and Sixty-one
Southern Alberta Jubilee Auditorium
Calgary, Alberta

Program

ACADEMIC PROCESSION

O CANADA

THE INVOCATION
The Very Reverend Monsignor John J. O'Brien

REPORT OF THE PRESIDENT TO CONVOCATION

REPORT OF PRINCIPAL M. G. TAYLOR

PRESENTATION OF THE ALUMNI GOLDEN JUBILEE AWARD
to
William Herbert Swift, B.A., B.Ed., M.A., Ph.D.
by Mr. H. Thomson

CONFERRING OF HONORARY DEGREE OF DOCTOR OF LAWS
Ian Nicholson McKinnon
Presented by Dean G. W. Govier

The Honorable Chief Justice Colin Campbell McLaurin
Presented by Dr. S. H. McCuaig, Q.C.

CONVOCATION ADDRESS
The Honorable Chief Justice Colin Campbell McLaurin

CONFERRING OF DEGREES
The Presentation
The Pledge
The Admission

PRESENTATION OF DIPLOMAS

GOD SAVE THE QUEEN

RECESSION

Music by the Regimental Band of the Lord Strathcona Horse (Royal Canadians) under the direction of Captain F. M. McLeod, C.D., by the kind permission of Lieut-Col. I. McD. Grant, D.S.O., C.D., Commanding Officer

Awards

The place mentioned after each prize winner's name is that in which his home is located.

UNDERGRADUATE AWARDS

THE FACULTY OF ARTS AND SCIENCE

The Edmonton Section of the Council of Jewish Women Scholarship
Hans Jurgen Reich, Edmonton

The Scholarship of the Board of Governors of the University in First Year Arts and Science
Gordon Charles Ball, Calgary (by reversion from James Howard Whittle, Edmonton)

The University Women's Club of Calgary Scholarship
Audrey Olein Jackson, Calgary

The James McCrie Douglas Memorial Scholarship
Anthony Bosch, Lethbridge

The Mary Cameron Douglas Memorial Scholarship
Shauna Margaret Murray, Edmonton

The Edmonton Lodge of B'nai B'rith Scholarship
Judith Ann Gould, Edmonton

The CHCT-TV (Calgary) Scholarships
Gordon Charles Ball, Calgary   Ronald Martin Clowes, Calgary

The Society of Arts and Science, Calgary, Scholarships
Michael Francis McCann, Calgary   Maureen Anne Mulholland, Calgary

The Great Books Awards
Kathleen Elizabeth Helmer, Calgary   Lorraine Gertrude Milne, Calgary

The R. L. King Memorial Scholarship in Journalism
Maurice Yacowar, Calgary

St. Hilda's Undergraduate Scholarship
Myrna Janet Williams, Calgary

The Friends of the University Bursaries in Arts and Science
Kyril T. Holden, Edmonton   Bastiaan C. Van Fraassen, Edmonton

THE FACULTY OF AGRICULTURE

The Scholarship of the Board of Governors of the University in First Year Agriculture
David Jonathan Puzey, Red Deer

The Loveseth Limited Scholarships in Agriculture
Gerald Marvin Coen, Bashaw   Arthur Olaf Olson, Cranford (by reversion
Stanley Allen Krause, Hilda   from Gordon Robert Collier, Penhold)
        David Jonathan Puzey, Red Deer

The Alberta Dairymen's Association Scholarship in Dairying
Milton Ashton Wriglesworth, Edmonton

The Alberta Dairymen's Association Summer Work Awards
Horace S. Baker, Edmonton   Randall Milford Nichols, Red Deer

The Canada Packers Scholarship
Robert Bertram Church, Balzac

The J. K. Sutherland Memorial Scholarship
Allan Wayne Anderson, Czar

The Canadian Feed Manufacturers' Association (Alberta Division) Scholarship
Ronald John Holmlund, Wetaskiwin

The Alberta Branch of the Canadian Seed Growers' Association Scholarship
Donald Edwald Harder, Rosemary

The Robert Gardiner Memorial Undergraduate Scholarship
Robert Bertram Church, Balzac

The Ketchum Prize in Animal Science
Ronald Edward Mathison, Dewberry

The Dan Baker Scholarships
Gerald M. Coen, Bashaw   Ronald Edward Mathison, Dewberry

The Douglas Farquhar Johnston Bursary in Agriculture
Allan Wayne Anderson, Czar

THE FACULTY OF COMMERCE

The Winspear, Hamilton, Anderson and Co. Scholarship
Stanley Sanderson, Calgary

The Hudson's Bay Company Scholarship in Commerce
Robert Mervyn Lewis, Grande Prairie

The Frederick Charles Manning Scholarship
Neil Ogilvie Henry, Edmonton

The Scholarship of the Western Daily Newspaper Advertising Managers' Association
James George Thoman, Edmonton

The Hudson's Bay Company Prize
Muriel Ann Coambs, Edmonton

The Institute of Chartered Accountants of Alberta Prize in Second Year Commerce
Robert Mervyn Lewis, Grande Prairie

The Clarkson Gordon and Co. Prize
Gordon Walter Flynn, Edmonton

The Business and Professional Women's Club, Edmonton Branch Scholarship in Commerce
Muriel Ann Coambs, Edmonton

The T. Eaton Company's Prizes in Commerce
In the second year: Robert Mervyn Lewis, Grande Prairie
In the first year: Rolf Gerhard Hattenhauer, Edmonton

The Riddell, Stead, Graham and Hutchison Award
Clifford Russell Thomas, Edmonton

The Friends of the University Bursary in Commerce
Rolf G. Hattenhauer, Edmonton

THE FACULTY OF DENTISTRY

The International College of Dentists (Canadian Section) Scholarship in Dentistry
Lawrence John Heppler, Edmonton

The Alberta Dental Association Scholarships
In the third year: Alexander Grigoruk, Edmonton
In the second year: Edward Clark, Edmonton
In the first year: Paul Duane Henderson, Edmonton

The Alberta Dental Association Prize in Second Year Dentistry
Robert Thomas Malpass, Nanaimo, British Columbia

The Calgary and District Dental Society Scholarship
Frank Lowell Peterson, Edmonton

The Edmonton Dental Society Prizes
In the third year: Jack Almquist Derbyshire, Edmonton
In the first year: William Robert Sproule, Vancouver, British Columbia

The Arthur D. Cumming Scholarships
Entering third year: Frank Lowell Peterson, Edmonton
Entering second year: Ervin Delmar Lietz, Edmonton
Entering first year: Ronald Frederick Ebdon, Edmonton

The Alberta Dental Association Scholarship in Pre-professional Year
Sam Masami Hoshizaki, Kelowna, British Columbia

The College of Dental Surgeons of Saskatchewan Scholarship
Robert Nelson Hicks, Vancouver, British Columbia

The Fred Stapells Scholarship in Dentistry
James Charles English, Edmonton

The Elizabeth and Wesley Haynes Prize
Kenneth Ray Nielsen, Edmonton

The Harry Ernest Bulyea Scholarship
Alexander Grigoruk, Edmonton

The Harry Alexander Gilchrist Scholarship
Jack Almquist Derbyshire, Edmonton

The Dental Faculty Class of 1950 Bursary
Edward Clark, Edmonton

The C. V. Mosby Book Awards
Edward Clark, Edmonton
Alexander Grigoruk, Edmonton
Malcolm Gary McRae, Port Coquitlam, British Columbia

The Dean's Scholarship in Dentistry
Gary Bruce Gibson, Burnaby, British Columbia

The Friends of the University Bursaries in Dentistry
Paul Duane Henderson, Edmonton
Robert F. C. Oswin, Edmonton

THE FACULTY OF EDUCATION

The Scholarship of the Board of Governors of the University in First Year Education
Betty Ann Rae, Medicine Hat

The University Scholarships in Education
Edith Pauline Bontus, Edmonton
Rachel Ann Martin, Duchess
Karen Margarethe Jenson, Ryley

The Edmonton Jewish Community Council Scholarship
Mary Ellena Davison, Edmonton

The P.E.O. Southern Alberta Scholarship in Education
Hilda Joyce Kunelius, Calgary

The Calgary Section of the Council of Jewish Women Scholarship
Thomas Perry McIntosh, Carbon

The First Year Scholarship of the Education Society of Edmonton
Betty Ann Rae, Medicine Hat

The N.O.M.A. Scholarship in Business Education
Ruth Etsuko Hashizume, Medicine Hat

The Olive M. Fisher Prize
Phyllis Audrey Morgan, Calgary

The Education Book Prize
Marilylle Pattison, Withrow

The Alberta Teachers' Association Scholarships to Students in Education
Joan Marie Coady, Edmonton
Margaret Donna Weir, Olds
Allan Edward Charles Shipton, Edmonton

Alberta Teachers' Association Scholarships to Teachers in the Field
Bette G Berent, Lethbridge
Thomas David Smith, Cereal

Smalley's Radio Ltd. Scholarship in Education
Stephen J. Baranek, Calgary

The Friends of the University Bursaries in Education
Betty Ann Rae, Medicine Hat
Leonard L. Rusnak, Glen Park

THE FACULTY OF ENGINEERING

The American Society for Metals Foundation Scholarship in Metallurgy
Robert Lawson Edgar, Edmonton

The American Society for Testing Materials Prize Awards
Roderick George Fujaros, Vegreville
Colin Douglas Laing, Edmonton
Ralph C. G Haas, Edmonton
Robert James Miller, Edmonton
John William Jensen, Bowden

The Scholarship of the Board of Governors of the University in First Year Engineering
Larry Wayne Toso, Calgary

The British America Paint Company Scholarship in Engineering
Robert Ridgeway Gilpin, Viking

The Canada Cement Company Scholarships
In Civil Engineering: Craig Evan Harrold, Edmonton
In Mechanical Engineering: John William Jensen, Bowden

The Chemical Institute of Canada Prize in Chemical Engineering
Harvie Andre, Edmonton

The C.I.S.C. Scholarship in Structural Engineering
Gordon Francis Anderson, Camrose

The A. Cristall Memorial Scholarships
Barry William Spence, Edmonton   Larry Wayne Toso, Calgary

The Dow Chemical Scholarship in Chemical Engineering
Harvie Andre, Edmonton

The Engineering Institute of Canada Prize for Third Year Students in Engineering
Malcolm William Wilson, Edmonton

The Gas Companies Undergraduate Scholarships in Engineering
Walter Clements Harrison, Medicine Hat
Roderick George Fujaros, Vegreville

The Johnston Testers Limited Scholarship
Larry Lorne McClennon, Redwater

The John Alexander McDougall Memorial Scholarship in Civil Engineering
Robert Francis Manuel, Camrose

The Schlumberger of Canada Scholarship
Douglas Neville Davis, Sedgewick

The H. R. Webb Memorial Scholarship
Cornelis Adriaan Muilwyk, Lethbridge

The John Wilcox Memorial Prize
Larry Lorne McClennon, Redwater

R. Angus (Alberta) Limited Scholarship in Highway Engineering
Stephen V. Benediktson, Edmonton

The California Standard Company Scholarship in Petroleum or Chemical Engineering
Malcolm William Wilson, Edmonton

The Sam J. Gorman Memorial Bursary in Engineering
Kwok-Fu Lam, Edmonton

The Douglas Farquhar Johnston Bursary in Engineering
Ralph Clarence Dorward, Viking

Peter Kiewit Sons Company of Canada Ltd. Scholarship
Norman Henry Neufeld, Granum

The James A. Lewis Engineering Scholarship in Petroleum Engineering
Deferred

The Mobil Oil of Canada, Ltd. Scholarship in Petroleum Engineering
Robert D. Bertram, Edmonton

The R.C.E. Memorial Scholarship
Not awarded

THE SCHOOL OF HOUSEHOLD ECONOMICS

The Mabel Patrick Scholarship
Elizabeth Vera Hempstock, Edmonton

The Florence Hallock Memorial Prize
Margaret Ann Stout, Edmonton

The Annie Gertrude Tory Scholarship
Elizabeth Vera Hempstock, Edmonton

The Gretta Shaw Simpson Memorial Prize
Leola Mae Calderwood, Edmonton

The Maria Isabel O'Connor Prize
Elizabeth Vera Hempstock, Edmonton

The Phylis Osborne McGachie Bursary in Household Economics
Patricia Marie Hyduk, Edmonton

The Friends of the University Bursary in Household Economics
Elizabeth Marilyn Barr, Edmonton

The Alberta Wheat Pool Scholarships in Household Economics
In the third year: Ruby Marie Butterwick, Brownfield
In the second year: Jean Marie Holmlund, Wetaskiwin

FACULTY OF LAW

The Scholarship of the Board of Governors of the University in First Year Law
Sheldon Mervin Chumir, Calgary

The Edmonton Lodge of B'nai B'rith Prize in Second Year Law
Clifton David O'Brien, Edmonton

The Sydney B. Woods Memorial Prize in Constitutional Law
Anton M. S. Melnyk, Edmonton

The William Gordon Egbert Prize in Administrative Law
Walter Shandro, Edmonton

The Carswell Prizes in the Faculty of Law
In the second year: Anton M. S. Melnyk, Edmonton
In the first year: Richard Douglas Tingle, Edmonton

THE FACULTY OF MEDICINE
General

The Research Fellowship of the College of Physicians and Surgeons of the Province of Alberta
Not awarded

The Scholarships of the College of Physicians and Surgeons of the Province of Alberta
In first year Proficiency (1st): Martin Hugh Atkinson, Calgary
In first year Proficiency (2nd): Mervyn Prisiart Williams, Edmonton
In second year Proficiency (1st): Elliott Asher Phillipson, Edmonton

In second year Proficiency (2nd): Edgar Garner King, Edmonton
In third year Proficiency (1st): Joseph Boyd Martin, Duchess
In third year Proficiency (2nd): Carol Ann Cowell, Calgary

The Dr. Allan Coates Rankin Award
Carol Ann Cowell, Calgary

Friends of the University Bursary in Medicine
Jerrold Nelson Finnie, Lowell, Massachusetts

Bacteriology

The Allan Coates Rankin Prize in Bateriology
Elliott Asher Phillipson, Edmonton

Medicine

The Prize of the Associated Hospitals of Alberta in Medicine
Joseph Boyd Martin, Duchess

The Prize in the History of Medicine
Not awarded

Obstetrics and Gynaecology

The Harrison Memorial Scholarship in Obstetrics and Gynaecology
James Richard Miller, Calgary

Physiology and Pharmacology

The Ardrey W. Downs Prize in Physiology
Thomas Henry Greidanus, Edmonton

The Nathan Browne Eddy Prize in Pharmacology
Jerrold Nelson Finnie, Lowell, Massachusetts

Surgery

The Prize of the Associated Hospitals of Alberta in Surgery
Joseph Shuster, Montreal

THE SCHOOL OF NURSING

The President's Gold Medal in Nursing and the Prize of the Chairman of the University Hospital Board
Elizabeth Mary Achtemichuk, Myrnam

The Prizes of the Board of Govrnors of the University for Nursing
For outstanding merit in nursing during the senior year:
Eimko Adachi, Coalhurst
For outstanding merit in the theory of Nursing during the senior year:
Bertha Kurylowich, Grimshaw
For outstanding merit in the practice of nursing during the senior year:
Elizabeth Mary Achtemichuk, Myrnam

The Edmonton Home Economics Association Prize
Doris Pearl Munroe, Innisfail

The Medical Staff of the University of Alberta Hospital
(a) *Strathcona Prize in Medicine*: Eleanor Margaret Gaychuk, Edmonton
(b) *Strathcona Prize in Obstetrics and Gynaecology*: Juane Elsie Mattson, Bellevue
(c) *Strathcona Prize in Psychiatry*: Vivian May Swischuk Bullen, Calgary
(d) *Strathcona Prize in Surgery*: Audrey Frances Duggan Tod, Edmonton

The Operating Room Nursing Prize
Lesley Marjorie R. B. Munroe, Swift Current, Saskatchewan

The Helen Smith Peters Memorial Prize
Emiko Adachi, Coalhurst

The Dr. A. C. McGugan Prize
Laurene Isabel Jickling, Provost

The Prize for Bedside Nursing
Virginia Ruth Thompson, Edmonton

The University of Alberta Hospital School of Nursing Alumnae Association Prize
Lois Vivian Kvisle, Innisfail

The Meta M. Hodge Memorial Prize
Elizabeth Mary Achtemichuk, Myrnam   Joyce Annabelle Cook, Edmonton

The Canadian Nurse Magazine Prize
Lorraine Caroline Dzwiniel, Edmonton   Else Ruth Barg, Brooks

The Women's Auxiliary of the University Hospital Prize
Marion Gladys McDonnell, Camrose   Barbara Luella Surbeck, Edmonton

The Friends of the University Bursary in Nursing
Elizabeth M. Achtemichuk, Myrnam

THE FACULTY OF PHARMACY

The Alberta Pharmaceutical Association Scholarships
In the second year: Rita Elly Pilger, Ohaton
In the first year: Gerald Herman Hirsch, Bawlf

The Alberta Pharmaceutical Prize in Dispensing
Edwin Wayne Howe, Edmonton

The M. J. Warner Scholarship
Anita Fongching Cheng, Edmonton

The Shipley Scholarship
Edwin Wayne Howe, Edmonton

The Calgary Ladies Pharmaceutical Auxiliary Prize
Beverley Anne Shewchuk, Edmonton

The Canadian Foundation for the Advancement of Pharmacy Scholarships
In the second year: Elsie Sandra Buxton, Mayerthorpe
Phyllis May Douglas, Brooks
In the first year: Ann Christine Careless, Edmonton
Sylvia Joan Van Haitsma, Camrose

The Charles E. Frosst Scholarship
Mary Ann Sheahan, St. Paul

The Charles Cummer Memorial Prize
Margaret Sue Colthorpe, Yellowknife, N.W.T.

The Burroughs Wellcome Bursary
Lorne Henry Baldwin, Calgary

The Merck Prizes
In Chemistry 230: Lorne Henry Baldwin, Calgary
In Chemistry 250: Gerald Herman Hirsch, Bawlf

The Drug Travellers of Alberta Bursary
Patricia Ann Sherbanuk, Edmonton

The Southern Alberta Pharmacy Bursary
Edward George Hunter, Lethbridge

THE SCHOOL OF PHYSICAL EDUCATION

The Edmonton Women's Branch of the Canadian Association for Health, Physical Education and Recreation
Esther Frances Segal, Edmonton

The J. K. Campbell and Associates Limited Scholarships
In the second year: Josephine Elizabeth Gozelny, Edmonton
In the first year: Albert Kenneth Dawson, Winnipeg Beach, Manitoba

The Edmonton Branch of the Canadian Association for Health, Physical Education and Recreation Bursaries
In the second year: Eunice Muriel Mattson, Medicine Hat
In the first year: Albert V. Carron, Edmonton

The Fruehauf Trailer Company of Canada Ltd. Scholarship
Peter J. Dooling, Edmonton

THE SCHOOL OF PHYSICAL THERAPY AND OCCUPATIONAL THERAPY

The Saskatchewan Physical Therapists Association Prize
Frances Ann Hayes, Calgary

The Physiotherapy Book Prize
Carolle Elizabeth Patterson, Edmonton

The Canadian Foundation for Poliomyelitis Bursaries
Bernice Isabelle Calvert, Edmonton   Jane Marie Vagt, Grande Prairie
Sandra Lynn Sundset, Bashaw

GENERAL AWARDS
(Open to more than one Faculty)

The Aluminum Company of Canada, Limited Scholarship
Hans Jurgen Reich, Edmonton

The Viscount Bennett Undergraduate Scholarships
Carol Ann Cowell, Calgary   Bevis Allan Ostermann, Bowness
William Richard Goddard, Calgary   John Bernard Ower, Calgary
David William Gussow, Calgary   Norma Joan Scott, Banff
Hilda Joyce Kunelius, Calgary   Dixon A. R. Thompson, Calgary
Richard J. W. Mansfield, Calgary

The William Asbury Buchanan Bursary
Anthony Bosch, Lethbridge

The California Standard Company Geophysical Scholarship
Gerritt T. F. R. Maureau, Edmonton

The Canadian Legion Scholarship in Physics or Electrical Engineering
Anton Willem Colijn, Calgary

The Canadian Legion Scholarship
Harold Robert Kane, Hillcrest

The Canadian Superior Bursaries
William Glen Johnston, Edmonton   Arthur Richard Sweet, Lacombe

The City of Calgary Undergraduate Scholarships
Phillip John Hadfield, Calgary   Michael Francis McCann, Calgary

The City of Edmonton Undergraduate Scholarships
Max Blake Burbank, Lethbridge   Gerald Leon Mayer, Edmonton
Kyril T. Holden, Edmonton   Gordon Theodore Sande, Edmonton

Cominco Undergraduate Scholarship
Ronald Martin Clowes, Calgary

The French Government Bursary
Not awarded

The German Academic Exchange Scholarship
Wendel Eugene Goodrich, Hardisty

The Glendale Kiwanettes of Calgary Scholarship
James H. Bjerring, Calgary

The P. Lawson Travel Ltd. Summer Scholarship
Kaaren M. Soby, Calgary

The Reg. Lister Trophy and Prize
Walter Clements Harrison, Medicine Hat

The Samuel J. McCoppen Bursaries
Dieter Kurt Buse, Barrhead   Lois M. Olsen, Bawlf
Arthur Wesley Cragg, Edmonton   Melvin W. Weber, Castor

The J. M. MacEachran Essay Prizes
First: John Ivan Marki, Edmonton
Second: Jerald B. Le Baron, Cardston
Third: Kendra Ann Aalgaard, Carmangay

The Kenneth William Moodie Scholarship
Anthony A. Fisher, Edmonton

The N.F.C.U.S. Interregional Exchange Scholarships
Elaine Stringam, Lethbridge   Patricia Ann Spence, Edmonton

The Northern Electric Undergraduate Scholarship
Cornelis Adriaan Muilwyk, Edmonton

The President's Scholarships
Ronald E. D. McClung, Edmonton   John Peter Unrau, Edmonton
Clifton David O'Brien, Edmonton

The Procter and Gamble Student Bursaries
Patrick George Collette, Calgary   Janos Rudolph Low, Edmonton
James Richard Connie, Edmonton   Frances E. Lozeron, Sexsmith
Neil William MacDonald, Crossfield   Rita Elly Pilger, Ohaton

The Renkenberger Scholarship
Earl Angus Jensen, Benalto

Smalley's Radio Ltd. Bursary in Arts and Science
Robin Patrick Rankine, Calgary

The Sunwapta Broadcasting Company Scholarship
Neil William MacDonald, Crossfield

The Robert Tegler Special Scholarship
Arthur Haviland Elford, Edmonton

The Women's Canadian Club of Edmonton Bursary
Helfried Werner Seliger, Edmonton

DEPARTMENTAL AWARDS

CHEMISTRY

The Chemical Institute of Canada Prize in Chemistry
Anthony Bosch, Lethbridge

The Lehmann Prize in Chemistry
Gordon Charles Ball, Calgary

The Paul Edward Macleod Memorial Prize in Chemistry
Michael Francis McCann, Calgary

The Fred H. Irwin Memorial Prize in Organic Chemistry
Anthony Bosch, Lethbridge and Nick Henry Werstiuk, Vilna (equal)

CLASSICS

The Ahepa Prize in Greek
Joan Ruth Sandilands, Edmonton

The Monica Jones Aamodt Prize in Classics in English 350
Joan Ruth Sandilands, Edmonton

ENGLISH

The Aiken Scholarship in English Language and Literature
John Peter Unrau, Edmonton

The Priscilla Hammond Memorial Scholarship in Honors English
Tiit Kodar, Edmonton

The James Patrick Folinsbee Memorial Prize
Not awarded

The Samuel Richard Hosford Memorial Prize
Ann Margaret Hardy, Edmonton

The Priscilla Hammond Memorial Prize in English 200
Jerald Bentley Le Baron, Cardston

The Calgary Herald Prize in English 240
Gary Donald Willis, Calgary

The Charles James Thompson Memorial Prize in English 220
Arthur Olaf Olson, Cranford

FINE ARTS

The Fuller Brush Company Scholarships in Art
John Robert McKee, Calgary   Ruth Leah Promislow, Edmonton

The University of Alberta Alumni Prize in Fine Arts
Margaret Rose Kitsco, Edmonton

The Lerner Scholarship in Drama
Karen Elizabeth Austin, Edmonton

The Elizabeth Sterling Haynes Scholarships in Drama
Karen Elizabeth Austin, Edmonton   Lloy Patricia Coutts, Edmonton

GEOGRAPHY

The Alberta Geography Prize
Jerald Bentley Le Baron, Cardston

GEOLOGY

The John A. Allan Memorial Scholarship in Geology
Gerald John Ryznar, Coleman

The Dome Petroleum Limited Scholarship
Gunter Kurt Muecke, Calgary

The Califorina Standard Company Geological Scholarship
Charles Ernest Johnson, Stettler

The Mobil Oil of Canada Ltd. Scholarship in Geology
Garry K. C. Clarke, Edmonton

HISTORY

The George Malcolm Smith Memorial Prize
Kurt Rees, Calgary

The History Club Prize
John Peter Unrau, Edmonton

The A. L. Burt Prize
Sheldon Mervin Chumir, Calgary

The Morden Heaton Long Prize
Ann Margaret Hardy, Edmonton

The Calgary Canadian Club History Prize
Donald Marvin Carr, Innisfail

The Hudson's Bay Company Scholarship
Ronald Gordon Seale, Edmonton

MATHEMATICS

The James McNish Chalmers Memorial Prize
Joe Thomas Roy Clarke, Edmonton

The Ernest Wilson Sheldon Memorial Prize in Mathematics
Wytze Brouwer, Edmonton

The Annie Fefferman Prize in Mathematics
Gordon Theordore Sande, Edmonton

The Mary Wyman Memorial Prize in Mathematics
Ronald Martin Clowes, Calgary

MODERN LANGUAGES

The Aristide Blais Prize in French
Doreen Gloria Nystrom, Blairmore

The Belgian Government Prize
Robert Knipe, Calgary

The Minister of Switzerland's Book Prize in French
Christian Prohom, Edmonton

The French Government Book Prize
William Scott Allison, Edmonton

The Beauchemin Prize in French
Audrey Olein Jackson, Calgary and Garry Donald Willis, Calgary (equal)

The German Government Prize
German 370: John Peter Unrau, Edmonton
German 450: Hermann Wilhelm Janzen, Edmonton

The Russian Literature Book Prizes
Russian 360: Gerald Leon Mayer, Edmonton
James Allen Radford, Calgary

The Canadian Legion, Norwood Branch, Scholarships in Ukrainian
William Douglas Kobluk, Delburne   Shirley Anne Yaremie, Andrew
Chrystyna Lukomska, Edmonton

The Ukrainian Catholic Women's Award in the Ukrainian Language
Daria Bohdana Gural, Radway

PHYSICS

The Society of Exploration Geophysicists Scholarships
Arnold A. Aylesworth, Calgary   Gerritt T. Maureau, Edmonton
William Hung Kan Lee, Edmonton

POLITICAL ECONOMY

The Men's Economics Club Prize
Rolf Gerhard Hattenhauer, Edmonton

FIRST CLASS STANDING AND HONOR PRIZES

In the following list of students those whose names are marked with an asterisk (*) have qualified for University of Alberta Honor Prizes. Those whose names are not so marked have qualified for University of Alberta First Class Standing Prizes.

Faculty of Arts and Science
Honors Courses

Third Year:
  Kenneth P. Bobrosky, Drumheller
  *Anthony Bosch, Lethbridge
  Adolf Buse, Barrhead
  Edward L. Dedio, Onoway
  Daniel De Vlieger, Edmonton
  Barry E. Galbraith, Edmonton
  David F. Goble, Waterton Park
  Robert Knipe, Calgary
  Tiit Kodar, Edmonton
  *Gerald L. Mayer, Edmonton
  Shauna M. Murray, Edmonton
  *James A. Radford, Calgary
  Helfried W. Seliger, Edmonton
  Nick H. Werstiuk, Vilna

Second Year:
- Morris L. Aizenman, Calgary
- *Frederick C. C. Andrews, Vulcan
- Patricia L. Baxter, Edmonton
- *Carl A. Blashko, Andrew
- Garry K. C. Clarke, Edmonton
- *Anton W Colijn, Calgary
- *Ronald S. Davis, Edmonton
- *Anthony A. Fisher, Edmonton
- Gerald F. Gabel, Medicine Hat
- *Judith A. Gould, Edmonton
- Ann M. Hardy, Edmonton
- *Richard J.W. Hodgson, Edmonton
- *Kyril T. Holden, Edmonton
- *Ronald E. D. McClung, Edmonton
- *George Mah Poy, Ponoka
- *Howard L. Malm, Vauxhall
- *Richard J. W. Mansfield, Calgary
- Gerritt T. F. R. Maureau, Edmonton
- *Gunter K. Muecke, Calgary
- John B. Ower, Calgary
- *Verner H. Paetkau, Coaldale
- *Kenneth Piers, Neerlandia
- Bente G. Rasmussen, Edmonton
- Ross A. Rudolph, Edmonton
- Donald G. Shaefer, Duffield
- Bastiaan C. Van Fraassen, Edmonton

General Courses

Second Year:
- *Heather J. Begg, Edmonton
- Murray S. Bogorus, Edmonton
- Donald M. Carr, Innisfail
- *David T. P. Dawson, Brant
- Ihor Dmytruk, Edmonton
- Charles D. Ferris, Edmonton
- Robert L. Fisk, Edmonton
- John G. Gokiert, Edmonton
- Marjorie E. Gold, Calgary
- Carole Grayson, Calgary
- Ronald C. Gregg, Edmonton
- Kenneth L. Hicken, Raymond
- Joe C. Lavery, Kelowna, B.C.
- John R. McKee, Calgary
- *Stuart H. Marston, Whitehorse, Y.T.
- Mattie S. Miller, Greenwood, B.C.
- Denis K. O'Gorman, Calgary
- David W. Penner, Calgary
- Dennis W. Schneck, Wetaskiwin
- John C. Stevens, Llyodminster
- *John P. Unrau, Edmonton
- Myrna J. Williams, Calgary

First Year:
- *William S. Allison, Edmonton
- *Gordon C. Ball, Calgary
- James H. Bjerring, Calgary
- Julia M. Breeze, Calgary
- Max Burbank, Lethbridge
- *Ronald M. Clowes, Calgary
- Peter Csontos, Daysland
- Eric D. Dandell, Red Deer
- Merete W. Frohn, Edmonton
- Michael W. Gray, Medicine Hat
- Frank B. Griblin, Calgary
- Gay M. Gullekson, Hanna
- Sam M. Hoshizaki, Kelowna, B.C.
- John Humphries, Edmonton
- Audrey O. Jackson, Calgary
- William R. Johnston, Calgary
- *Harold R. Kane, Hillcrest
- Kimio Kinoshita, Edmonton
- Willy A. G. Krynen, Edmonton
- Jerald B. Le Baron, Cardston
- Randy K. Lomnes, Wetaskiwin
- *Michael F. McCann, Calgary
- E. Preston Manning, Edmonton
- *Edward A. Meighen, Edmonton
- Ian H. Pitfield, Edmonton
- Donald G. Rawlinson, Edmonton
- Buren R. Ree, Bentley
- *Hans J. Reich, Edmonton
- *Gordon T. Sande, Edmonton
- Dixon A. R. Thompson, Calgary
- *James H. Whittle, Edmonton
- Gary D. Willis, Calgary

Faculty of Agriculture

Third Year:
- Allan W. Anderson, Czar
- *Robert B. Church, Balzac
- Lawrence W. Copithorne, Cochrane

Second Year:
- Gerald M. Coen, Bashaw

First Year:
- Stanely A. Krause, Hilda
- David J. Puzey, Red Deer

Faculty of Commerce

Second Year:
- Robert M. Lewis, Grande Prairie

First Year:
- None

Faculty of Dentistry

Third Year:
- Alexander Grigoruk, Edmonton

Second Year:
   Edward Clark, Edmonton   Frank L. Peterson, Edmonton

First Year:
   Paul D. Henderson, Edmonton

Faculty of Education

Fourth Year:
   Emerson R. Shantz, Didsbury   Dorothy F. Tetley, Red Deer
   Joseph A. Stevenson, Edmonton   Gisela E. A. Wulff, Edmonton

Third Year:
   Halia K. Boychuk, Ashmont   *Marilylle Pattison, Withrow
   Elsie R. Daintith, Camrose   Victor Rempel, Bowness
   Florence E. Irwin, Edmonton

Second Year:
   Vera I. Bracken, Medicine Hat   Mary Lobay, Edmonton
   Mary E. Davison, Edmonton   Elsie M. McRoberts, Irma
   Lillian J. Gillespie, Edmonton   Manfred H. Rupp, Edmonton
   Edward K. Hostetler, Tofield   Norma J. Scott, Banff
   Lawrence A. Kwiatowski, Warburg   Nick Tkachuk, Ranfurly

First Year:
   Alan R. Bromley, Two Hills   James O. Ramsay, Edmonton
   Agnes R. Hebert, Edmonton   Maxine L. Runions, Edmonton
   Betty A. Rae, Medicine Hat   Leonard L. Rusnak, Glen Park

Junior E:
   Patricia R. Bolinger, Gleichen

Faculty of Engineering

Third Year:
   Gordon F. Anderson, Camrose   Bertram C. McInnis, Turner Velley
   *Harvie Andre, Edmonton   John C. Moldon, Winnipeg, Man.
   *Lorne E. Carlson, Edmonton   *Robert F. Manuel, Edson
   *Douglas N. Davis, Sedgewick   Arie Reedyk, Lethbridge
   Ralph C. Dorward, Viking   Malcolm W. Wilson, Edmonton
   Leonard Forbes, Hythe

Second Year:
   *Norman L. Arrison, Bassano   *Kwok-Fu Lam, Hong Kong
   Clifford S. Ashley, Edmonton   Robert J. Miller, Edmonton
   Michael R. Day, Edmonton   Bevis A. Osterman, Bowness
   Bruce T. Dippie, Edmonton   John H. Pankiw, Whilelaw
   Marvyn G. Faulkner, Mayerthorpe   Frederick A. Seyer, Botha
   *Roderick G. Fujaros, Vegreville   Thomas E. Siddon, Drumheller
   William R. Goddard, Calgary   *Robert A. Smith, Lethbridge
   Lloyd J. Griffiths, Edmonton   *Theodore D. Strashok, Edmonton
   *Walter C. Harrison, Medicine Hat   Charles G. Sutton, Edmonton
   Craig E. Harrold, Edmonton   Kent R. Tiedemann, Tofield
   Donald E. Holte, Edmonton   *Joseph D. Wright, Three Hills
   John W. Jensen, Bowden

First Year:
   Wesley Abel, Calmar   Stacey D. Jarvin, Rocky Mtn. House
   David R. A. Budney, Lamont   Hugh D. Jones, Calgary
   *Alvin R. Enns, Saskatoon, Sask.   *Cornelis A. Muilwyk, Lethbridge
   Clemens R. Feldmeyer, Calgary   *Roger A. Pretty, Edmonton
   *Robert R. Gilpin, Viking   *David Routledge, Edmonton
   *Jeffry J. Hakeman, Edmonton   *Larry W. Toso, Calgary
   *David E. Holroyd, Rocky Mtn. House   *Harvey L. Werstiuk, Glendon
   *Peter H. Holst, Edmonton   Robert Wolfe, Edmonton
   William W. Irwin, Calgary

School of Household Economics

Second Year:
   *Elizabeth V. Hempstock, Edmonton

First Year:
   Elizabeth M. Barr, Edmonton   Margaret A. Stout, Edmonton

Faculty of Law

Second Year:
   Clifton D. O'Brien, Edmonton

First Year:
   Sheldon M. Chumir, Calgary

Faculty of Medicine

Third Year:
   Carol A. Cowell, Calgary   Joseph B. Martin, Duchess
   Morris Davidman, Calgary

Second Year:
   Edgar G. King, Edmonton   Elliott A. Phillipson, Edmonton

First Year:
   None

School of Nursing

First Year:
   None

Faculty of Pharmacy

Second Year:
   Edwin W. Howe, Edmonton   Mary A. Sheahan, St. Paul

First Year:
   *Lorne H. Baldwin, Calgary   Patricia A. Sherbanuk, Edmonton

School of Physical Education

Second Year:
   None

First Year:
   None

School of Physical Therapy and Occupational Therapy

Second Year:
   None

First Year:
   None

EXTRA-CURRICULAR AWARDS

Lorne Calhoun Memorial Award
Peter S. Hyndman, Edmonton

Block A Club Athletic Scholarship
Kenneth Ray Nielsen, Edmonton

THE PROVINCE OF ALBERTA UNDERGRADUATE SCHOLARSHIPS
(The Queen Elizabeth Education Scholarship Fund)

Wesley Abel, Calmar
Elizabeth Achtemichuk, Myrnam
William S. Allison, Edmonton
Frederick C. Andrews, Vulcan
Gordon F. Anderson, Camrose
Harvie Andre, Edmonton
Raymond H. Archer, Edmonton
Norman L. Arrison, Bassano
Martin H. Atkinson, Calgary
James P. Aunger, Stettler
Gordon C. Ball, Calgary
Elizabeth M. Barr, Edmonton
Patricia L. Baxter, Edmonton
Gerald F. Bennett, Spring Coulee
Heather J. Begg, Edmonton
John B. Bell, Marwayne
Stephan V. Benediktson, Edmonton
Lawrence M. Bezeau, Lethbridge
Larry J. Bilan, Tofield
James H. Bjerring, Calgary
John A. Blackett, Edmonton
Carl A. Blashko, Andrew
Kenneth P. Bobrosky, Drumheller
Murray S. Borgorus, Edmonton

Anthony Bosch, Lethbridge
Halia K. Boychuk, Ashmont
Jean C. Boyd, Calgary
Julia M. Breeze, Calgary
Walter H. Breitkreuz, Edmonton
Alan R. Bromley, Three Hills
Gary A. Brown, Calgary
Wytze Brouwer, Edmonton
Bonnie N. Bryans, Fort Sask.
Donald M. Buchanan, Edmonton
David R. A. Budney, Lamont
Max B. Burbank, Lethbridge
John G. Burch, Edmonton
Sara E. Burke, Edmonton
Adolf Buse, Barrhead
Elsie S. Buxton, Mayerthorpe
Charles D. Cairns, Calgary
Leola M. Calderwood, Edmonton
Clarence E. Capjack, Elk Point
Barry L. Caplan, Calgary
Ann C. Careless, Edmonton
Lorne E. Carlson, Edmonton
Donald M. Carr, Innisfail
Nora I. Chell, Fort MacLeod
Walter J. Chudobiak, Dalroy
Robert B. Church, Balzac
Edward Clark, Edmonton
√ Garry K. C. Clarke, Edmonton
Joe T. R. Clarke, Edmonton
Arthur L. Close, Edmonton
Ronald M. Clowes, Calgary
Joan M. Coady, Cardston
Gerald M. Coen, Bashaw
Anton W. Colijn, Calgary
Libby M. Cotsman, Edmonton
Robert J. Coulson, Nanton
Carol A. Cowell, Calgary
Peter Csontos, Edmonton
Eric D. Dandell, Red Deer
Karen A. Davey, Calgary
Michael Davidman, Calgary
Morris Davidman, Calgary
Douglas N. Davis, Sedgewick
Ronald S. Davis, Edmonton
David T. P. Dawson, Brant
Michael R. Day, Edmonton
Melvin de Levie, Edmonton
Jack A. Derbyshire, Edmonton
Stanley R. Didrikson, Wetaskiwin
Ralph C. Dorward, Viking
Donald M. Dublenko, Leduc
Donald G. Engelstad, Edmonton
Alvin R. Enns, Edmonton
Edda Ensslen, Edmonton
William W. Evans, Brownvale
David J. Fairholm, Pibroch
Marvyn G. Faulkner, Mayerthorpe
C. David Ferris, Edmonton
Jerrold N. Finnie, Edmonton
Anthony A. Fisher, Edmonton
R. Leighton Fisk, Edmonton
Leonard Forbes, Hythe
Robert M. W. Frederking, Vegreville
Kevin A. French, Edmonton
Peter H. Fuchs, Edmonton
Roderick G. Fujaros, Vegreville

Fred Y. Fujiwara, Brant
Robert Gaarlandt, Edmonton
Gerald F. Gabel, Medicine Hat
David R. Galbraith, Edmonton
Barry E. Galbraith, Edmonton
Lillian J. Gillespie, Edmonton
John R. Gallimore, Edmonton
Murray T. Gibson, Calgary
Robert R. Gilpin, Viking
David F. Goble, Waterton Park
William R. Goddard, Calgary
John G. Gokiert, Edmonton
Robert F. Gordon, Edmonton
Michael W. Gray, Medicine Hat
Carole M. Grayson, Calgary
Ronald C. Gregg, Edmonton
Lloyd J. Griffiths, Edmonton
Alexander Grigoruk, Edmonton
Helen R. H. Grout, Medicine Hat
David A. Grove, Edmonton
Gay M. Gullekson, Hanna
David W. Gussow, Calgary
Philip J. Hadfield, Calgary
Jeffrey J. Hakeman, Edmonton
Donald E. Harder, Rosemary
Nolan P. Haring, Stettler
Craig E. Harrold, Edmonton
Rolf G. Hattenhauer, Edmonton
Agnes R. Hebert, Calgary
Robert L. Hemmings, Edmonton
Elizabeth V. Hempstock, Edmonton
Paul D. Henderson, Edmonton
Neil O. Henry, Edmonton
Lawrence J. Heppler, Lethbridge
Kenneth L. Hicken, Edmonton
Gerald H. Hirsch, Bawlf
Richard J. W. Hodgson, Edmonton
Joanne M. Hohol, Barrhead
Kyril T. Holden, Edmonton
W. John Hollingsworth, Edmonton
David E. Holroyd, Rocky Mtn. House
Peter H. Holst, Edmonton
Donald E. Holte, Edmonton
Edward K. Hostetler, Tofield
Edwin W. Howe, Edmonton
Margaret Humphreys, Edmonton
John Humphreys, Edmonton
Marilyn A. Hutchinson, Edmonton
Patricia A. Hyduk, Edmonton
√ Peter S. Hyndman, Edmonton
W. Wayne Irwin, Calgary
Audrey O. Jackson, Calgary
Barbara A. Jackson, Edmonton
Juergen Jahn, Calgary
Stacey D. Jarvin, Rocky Mtn. House
John K. Jenkins, Edmonton
John W. Jensen, Bowden
Karen M. Jensen, Ryley
Laurence D. Jewell, Athabasca
Charles G. Johnson, Edmonton
Ronald M. Kachman, Strome
Harold R. Kane, Hillcrest
Harold J. Katzin, Edmonton
Patrick C. Kelly, Lake Majeau
Alec J. K. Keylock, Calgary
Kimio Kinoshita, Edmonton

Robert Knipe, Calgary
Tiit Kodar, Edmonton
Eileen J. Kristensen, Ponoka
Willy A. G. Krynen, Edmonton
Lorne A. Kuehn, Winterburn
Joseph Kuzmiski, Oyen
Sylvia A. J. Kuzyk, Bon Accord
Marilyn G. Lauber, Dodds
J. Bentley Le Baron, Cardston
Randy K. Lomnes, Wetaskiwin
Michael F. McCann, Calgary
Larry L. McClennon, Redwater
Ronald E. D. McClung, Edmonton
Lyall R. McCurdy, Edmonton
Ian R. McDonald, Edmonton
John R. McKee, Calgary
Alan R. MacLeod, Edmonton
Howard L. Malm, Vauxhall
✓ E. Preston Manning, Edmonton
Stuart H. Marston, Edmonton
Glen E. Martin, Lacombe
Joseph B. Martin, Edmonton
Lennart H. Marx, High Prairie
Richard J. W. Mansfield, Calgary
Robert F. Manuel, Edson
Gerritt T. F. R. Maureau, Edmonton
Gerald L. Mayer, Edmonton
Howard A. Meger, Heinsburg
Edward A. Meighen, Edmonton
Anton M. S. Melnyk, Edmonton
Mattie S. Miller, Edmonton
Robert J. Miller, Edmonton
John C. Moldon, Calgary
Robert H. Morse, Irama
Gunter K. Muecke, Calgary
Cornelis A. Muilwyk, Edmonton
Gordon Nicholson, Vegreville
Anthony F. Nield, Edmonton
Dale E. Niwa, Acadia Valley
Ross J. Norstrom, Edmonton
Doreen G. Nystrom, Blairmore
Clifton D. B. O'Brien, Edmonton
Herbert L. O'Brien, Edmonton
Larry W. Olesen, Edmonton
Arthur O. Olson, Cranford
John W. Olson, Red Deer
Bevis A. Ostermann, Bowness
Neva Y. Otsuka, Raymond
John B. Ower, Calgary
John H. Pankiw, Edmonton
Leslie J. Papp, Delburne
David W. Penner, Calgary
Carin E. Person, Edmonton
Frank L. Peterson, Edmonton
Robert L. Peterson, Strathmore
Eliot A. Phillipson, Edmonton
Kenneth Piers, Neerlandia
✓ Ian H. Pitfield, Edmonton
Roger A. Pretty, Edmonton
Carol J. Price, Picture Butte
John P. Quine, Edmonton
James A. Radford, Edmonton
Betty A. Rae, Medicine Hat
Annetta J. Ramsey, Edmonton
James O. Ramsay, Edmonton
Per G. Rasmussen, Edmonton

Donald G. Rawlinson, Edmonton
Stanley E. Read, Bashaw
Buren R. Ree, Bentley
Arie Reedyk, Lethbridge
Kurt Rees, Calgary
Hans J. Reich, Edmonton
Guenter W. Riedel, Fort MacLeod
Lorence N. Rempel, Edmonton
David Routledge, Edmonton
Ross A. Rudolph, Edmonton
Richard W. Rumpel, Edmonton
Maxine L. Runions, Edmonton
Donna L. Rusnak, Edmonton
Leondard L. Rusnak, Glen Park
Kendel J. Rust, Brooks
Gordon T. Sande, Edmonton
Joan R. Sandilands, Edmonton
Brian A. Saunders, Red Deer
Carol A. Schisseler, Medicine Hat
Dennis W. Schneck, Wetaskiwin
Ronald G. Seale, Edmonton
Eda M. Seeman, Lethbridge
Helfried W. Seliger, Edmonton
Maurice J. Seneshen, Edmonton
Lynn Sereda, Edmonton
William M. Sereda, Edmonton
Frederick A. Seyer, Botha
Carol J. Shantz, Didsbury
Patricia A. Sherbanuk, Edmonton
Edward E. Shubert, Edmonton
Thomas E. Siddon, Drumheller
Marilynn C. Smith, Edmonton
Percy A. Smith, Edmonton
Robert A. Smith, Lethbridge
Richard E. Snyder, Cowley
Barry W. Spence, Edmonton
John H. Sprague, Edmonton
Karl H. Stevens, Spruce Grove
Harry S. Stinchcombe, Edmonton
Margaret A. Stout, Edmonton
Theodore D. Strashok, Edmonton
William Y. Svrcek, Picture Butte
Ernest S. Takacs, Edmonton
Delia A. Tetreau, Vegreville
Dixon A. R. Thompson, Calgary
Donald H. Timbres, Edmonton
Guy J. Tourigny, Edmonton
John P. Unrau, Edmonton
Bastiaan V. van Fraassen, Edmonton
Sylvia J. Van Haitsma, Camrose
Frances H. Van Sant, Calgary
James H. Vance, Raymond
Jan F. Vaneldik, Edmonton
Curtis F. Vail, Edmonton
Maynard Vollan, Edmonton
John S. Watts, Calgary
Leonard N. Wedman, Calmar
Harvey L. Werstiuk, Glendon
Nick H. Werstiuk, Vilna
Ole Westby, Edmonton
Brian A. Westcott, Ralston
Florence R. Whiteley, Edmonton
J. Michael Whitla, Edmonton
Henry A. Wiebe, Coaldale
Leona J. Wiebe, Coaldale
Rose Marie Wilinski, Edmonton

Gary D. Willis, Calgary
Myrna J. Williams, Calgary
Mervyn P. Williams, Edmonton
Malcolm W. Wilson, Edmonton
Edna L. Winkelaar, Edmonton
Robert Wolfe, Edmonton

Milton A. Wriglesworth, Edmonton
James G. Wright, Rimbey
Jack F. Yasayko, Edmonton
Harold H. Yasayko, Edmonton
Dale R. Younge, Mannville

GRADUATE AWARDS

(The following graduate awards have been made since the publication of the Convocation Program for May, 1961.)

The following awards are open in two or more Fields of Study:

The Cominco Fellowship
Leslie Michael Lavkulich, Lethbridge

The Zonta Club of Edmonton Scholarship
Nadia Hochachka, Edmonton

The following awards are open in one Field of Study only:

EDUCATION

The John Walker Barnett Scholarship in Education
Lloyd Wilbert West, Edmonton

The Milton Ezra LaZerte Scholarship in Education
Kathlyn Benger, Edmonton

The Du Pont Company of Canada Limited Scholarship (1961)
Alexander James Dawson, Edmonton

ENGINEERING

The California Standard Company Graduate Fellowship in Petroleum Engineering
Ralph Elborne Hughes, Edmonton

GEOLOGY OR GEOPHYSICS

The California Standard Company Graduate Fellowship in Geology or Geophysics
Reginald Edward Griffiths, Calgary

PHARMACY

The Warner-Lambert Graduate Fellowship
Raymond Dennis Magus, Edmonton

The Pfizer Research Scholarship
M. N. Sukumaran Nayar, India

PHYSICAL EDUCATION

The Fruehauf Trailer Company of Canada Ltd. Scholarship
James L. Hodgson, Edmonton

MATRICULATION AWARDS

University of Alberta Matriculation Scholarships

Linda Margaret Hutchinson, Fort Sask.
Lavera Mae Sawatzky, Westlock
Sharon Ruth Widlake, Galahad

University of Alberta War Memorial Scholarships

Brian Adair Glanfield, Hoadley
Diane Helen Parsons, Westlock
Larry Virgel Pimm, Amisk
Aubrey James Tingle, Edmonton
Richard Dwight Williams, Calgary

The Alberta Hotel Association Scholarships

Peace River Zone

Annie M. Chop, Dixonville
Hazel G. McEachern, Hythe
Jane Shemanko, Rycroft

Northern Zone

Leondard G. Basaraba, Edson
Judith A. Buchanan, Athabasca
Cecelia A. Burak, Norma
Irene R. Chmilar, Vilna
Edward Chwyl, Andrew
Dianne E. Gerun, Perryvale
Alexander I. Gibson, Westlock
Norbert Goebel, Fort Sask.
Marilyn M. Gregorwich, Westlock
Marcia A. Hagen, Clandonald
Edward M. Holthe, Elk Point
Linda M. Hutchinson, Fort Sask.
Thomas A. Landsman, Duffield
Linda M. Libbey, Redwater
Carol A. Madsen, Mayerthorpe
Danny M. Mascaluk, Willingdon
Virginia Medwid, Grassland
Marilyn J. Olsen, St. Paul
Elizabeth Poss, Thorhild
Roland E. Theroux, Bonnyville

Central Zone

Barbara J. Erickson, Eckville
Sharon L. Ericson, Wetaskiwin
Robert J. Fairbairn, Vegreville
Robert G. Foulon, Camrose
Gilbert A. Schultz, Hay Lakes
Richard P. Terlesky, Buford
Vivien M. Tillett, Vermilion
Job W. Vander-Wekken, Lacombe
Dorothy J. M. Yarema, Holden
Ronald D. Yarmchuk, Red Deer

Special Awards

Patricia G. Smith, Hayter

Southern Zone

William J. Atwood, Lethbridge
James L. Fisher, Raymond
Gordon J. Fitzpatrick, Drumheller
Beverley M. Gietz, Pincher Creek
Darlene M. King, Three Hills
Eileen O. M. Kosior, Black Diamond
Carol E. A. Lencucha, Blairmore
Robert G. Moore, Seebe
Tomeko Nakagawa, Welling
Sandra E. Scheierman, Vulcan

Edmonton City

Leslie K. W. Brygadyr, Edmonton
Kenneth C. Dewar, Edmonton
Heinz E. Frank, Edmonton
Myron Goldberg, Edmonton
William Laskowski, Edmonton
Kathleen E. Nichols, Edmonton
Peter J. Puchyr, Edmonton
Elisabeth Smit, Edmonton
David Walden, Edmonton
Marie E. Zenko, Edmonton

Special Award

Herbert Bischof, Edmonton

Calgary City

John L. Broad, Calgary
Allan T. J. Connery, Calgary
Colleen L. Dagnall, Calgary
David J. Hall, Calgary
Hannelore Kermer, Calgary
Allister J. McPherson, Calgary
Frederick D. Parker, Calgary
John T. Price, Calgary
Gwyneth M. Roberts, Calgary
Rosina Urch, Calgary

The Alberta Institute of Agrologists Scholarship in First Year Agriculture

Joseph V. I. Noel, St. Lina

The Alberta Wheat Pool Bursaries for Rural Students

Marian Joan Dey, Ardrossan
Brian Stuart Ekstrom, Balzac

The Alberta Motor Transport Association Bursaries

Myron Goldberg, Edmonton
Judith Lynne Kales, Edmonton
Sharon Ruth Widlake, Galahad

The Viscount Bennett Matriculation Scholarships
Randy Perry Dyck, Balzac
Allan Richard Fabris, Calgary
Brian Douglas Warrack, Calgary

The Calgary and District Retail Druggists' Association Scholarship
Brian William Draper, Taber

The Calgary Men's Canadian Club Scholarship
Robert Barry Dunbar, Calgary

The Calgary Herald Scholarship in Chemical and Petroleum Engineering
Ronald Eugene Nieman, Calgary

The Calgary Herald Scholarship in Classics
John Norman Wood, Calgary

The Calgary Life Underwriters Association Bursary
Hans Joachim Brown, Calgary

The Canadian Bechtel Limited Bursary in Engineering
Eric Henry Shelton, Edmonton

The Canadian Foundation for the Advancement of Pharmacy Bursary
Joan Isabel Neilson, Onoway

The Canadian Foundation for Poliomyelitis and Rehabilition (Alberta Chapter) Bursaries
Marlene R. Campbell, Calgary
Urania Caroline Dong, Lethbridge
Marjorie Ellen Nant, Provost

The Canadian Freightways Motor Transport Matriculation Bursary
Colin Michael Evans, Cowley

The Canadian-Marietta Customers Scholarship
Brian Douglas Sykes, Edmonton

Canadian Mathematical Congress Scholarships in Mathematics
Canadian Mathematical Congress Scholarship
Egbert Krikke, Lacombe

Nickle Foundation Scholarship
Robert W. Taylor, Magrath

Haddin, Davis and Brown Ltd. Scholarship
Brian D. Sykes, Edmonton

Canadian Society of Exploration Geophysicists Matriculation Scholarship in Education
Garry Wayne Nyrose, Calgary

The Canadian Superior Bursary
Philip Golden Lister, Edmonton

The Cardston Pharmacy Scholarship
Joan Isabel Neilson, Onoway

The City of Calgary Matriculation Scholarships
Hans Brown, Calgary
Mavis Irene Dunford, Calgary
Donald Lloyd Evans, Calgary
Richard Allen Gusella, Calgary
Lorne Arthur Klippert, Calgary
Helmut Schwachhofer, Calgary

The City of Edmonton Matriculation Scholarships
Heinz Erich Frank, Edmonton
Robert Edward Miller, Edmonton

The City of Medicine Hat Scholarship
Anne Lucille Koenig, Medicine Hat

The Civil Service of Alberta Bursaries

John R. Caron, Edmonton
Michael W. Edwards, Edmonton
Patricia L. Harvie, Edmonton
Agnes E. Hereford, Coleman
Donna L. Roxburgh, Edmonton

Katheleen M. Savage, Ponoka
Vivian P. Smith, Olds
Phyllis I. Wickie, Calgary
Elaine E. Wuetherick, Edmonton

Cominco Undergraduate Scholarships

George Frederick Calow, Calgary   John William Gee, Edmonton

The E.I.C. Wives Club of Calgary Scholarship
Ola S. Juvkam-Wold, Calgary

Edmonton Civic Employees' Bursaries

David W. Darling, Edmonton
Robert W. Engley, Edmonton
Donald J. Keech, Edmonton

Robert B. Matheson, Edmonton
Sonja G. Procter, Edmonton
Edward J. Reid, Edmonton

The Edmonton and District Druggists' Association Scholarship
Vera Mary Kotylak, Edmonton

The Friends of the University Matriculation Bursaries

Colin Michael Evans, Cowley   Dianne Elizabeth Gerun, Perryvale

General Motors Scholarship
Eike Henner W. Kluge, Calgary

The C. O. Hicks Matriculation Scholarship
James R. G. Gardiner, Edmonton

The Imperial Order Daughters of the Empire Matriculation Bursaries

Brian A. Glanfield, Hoadley
Margaret J. Thompson, Edmonton

Lorna E. Woolfe, Beaverlodge
Ralph G. Yarstad, Taber

The Imperial Order Daughters of the Empire Second War Memorial Matriculation Scholarships

John Andrew Brook, Edmonton
Wendie M. Hodge, Calgary

Lisa Susan Southern, Scandia
Aubrey J. Tingle, Edmonton

The Imperial Order Daughters of the Empire Lethbridge Municipal Chapter Bursary
Gordon Holt, Lethbridge

The Inco Scholarship
Ellen Ann Sprinkle, Monarch

The Interfraternity Council Scholarship
James R. G. Gardiner, Edmonton

The L. T. Melton Real Estate Limited Scholarship
Morris Benjamin Draper, Edmonton

The North Calgary Business and Professional Women's Club Bursary
Marjorie E. Keller, Calgary

Petrofina Group Western Canada Scholarship
The Canadian Fina Scholarship
Ronald Allen Winter, Edmonton

St. Hilda's Scholarships

Anne Cara Matthews, Calgary Patricia Joan Wales, Calgary

The Robert David Sinclair Scholarship in First Year Agriculture
Keith Douglas Bresee, Ponoka

The Steel Company of Canada Bursary
Katherine Mary Stenger, Warburg

The Robert Tegler Matriculation Scholarships

Brian Gray Crummy, Edmonton √ Judith Lynne Kales, Edmonton
James R. G. Gardiner, Edmonton

The Robert Tegler Special Scholarship
Andrew Mark Stevens, Lloydminster

The University Women's Club of Edmonton Scholarship
Ellen Ann Sprinkle, Monarch

The Union Carbide Scholarship
Earl Roger Spady, Alliance

The Hon. W. C. Woodward Scholarship
John Douglas Mulholland, Edmonton

University of Alberta Honor Prizes

Glen S. Aikenhead, Calgary
Hans J. Brown, Calgary
Allan T. J. Connery, Calgary
Ulrike E. Conradi, Edmonton
Brian G. Crummy, Edmonton
Lynda J. Cunliffe, Edmonton
Kenneth R. Dubeta, Edmonton
Mavis I. Dunford, Calgary
Randy P. Dyck, Balzac
Christine E. Eder, Edmonton
Donald L. Evans, Calgary
Allan R. Fabris, Calgary
Maureen G. Fiddes, Gleichen
James L. Fisher, Raymond
Gordon J. Fitzpatrick, Drumheller
Robert G. Foulon, Camrose
Geoffrey G. Frank, Edmonton
Patricia E. Gander, Edmonton
James R. G. Gardiner, Edmonton
Beverley M. Gietz, Pincher Creek
Myron Goldberg, Edmonton
John C. E. Greene, Edmonton
Elaine Dorothy Hughes, High River
Linda M. Hutchinson, Fort Sask.
Dennis W. Jirsch, Edmonton
Ralph G. Jorstad, Taber
Judith L. Kales, Edmonton
Hannelore Kermer, Calgary
Lorne A. Klippert, Calgary
Eike-Henner W. Kluge, Calgary
Eileen O. M. Kosior, BlackDiamond
Egbert H. Krikke, Lacombe
Philip G. Lister, Edmonton
Carol A. Madsen, Mayerthorpe
Virginia Medwid, Grassland
Robert E. Miller, Edmonton
James M. Moscovich, Lethbridge
Douglas J. Mulholland, Edmonton
Tomeko Nakagawa, Welling
Mary Anne Nichols, Castor
Kathleen E. Nicholls, Edmonton
Garry W. Nyrose, Calgary
Vera A. Pederson, Hughenden
Peter J. Puchyr, Edmonton
Ilona A. Raycheba, Edmonton
Judith E. Reeves, Edmonton
Lavera M. Sawatzky, Westlock
Sandra E. Scheierman, Vulcan
Helmut Schwachhofer, Calgary
Dorothea H. Schwanke, Holden
M. Elaine Sereda, Edmonton
Eric H. Shelton, Edmonton
Edward E. Simbalist, Edmonton
Sheila N. Smeltzer, Edmonton
Earl R. Spady, Alliance
Ellen A. Sprinkle, Monarch
Katherine M. Stenger, Warburg
Donald E. Stephens, Red Deer
Edith J. Stilwell, Calgary
Brian D. Sykes, Jasper Place
Aubrey J. Tingle, Edmonton
John Van Dyke, Edmonton
David Walden, Edmonton
Patricia J. Wales, Calgary
Brian D. Warrack, Calgary
Hans Wehrfritz, Edmonton
Kenneth Welsh, Milk River
Sharon R. Widlake, Calgary
Marie E. Zenko, Edmonton

PROVINCE OF ALBERTA MATRICULATION SCHOLARSHIPS
(The Queen Elizabeth Education Scholarship Fund)

Susannea M. Abzinger, Calgary
Glen S. Aikenhead, Calgary
Israel Aizenman, Calgary
Dennis B. Alfke, Lacombe
Robert E. Allin, Edmonton
Dale F. Anderson, Grande Prairie
Mora L. Arthur, Calgary
William J. Atwood, Lethbridge
Caroline E. Baker, Edmonton
Marlene M. Bakkan, Calgary
Mary J. Batiuk, Edmonton
Mary L. Battell, Edmonton
Laurence E. Becker, Edmonton
Carol A. Bergquist, Bawlf
Henry L. Bertram, Calgary
Denise E. Biles, Calgary
Margaret J. Bonner, Edmonton
Edith L. Bioletti, Red Deer
John David Birrell, Calgary
Harold A. Bjorge, Bawlf
Morris Bodnar, Willingdon
Nancy E. Bowen, Edmonton
William W. Brand, Edmonton
Joyce A. Bredo, Edmonton
Wendy M. Brinsmead, Camrose
John Lewellyn Broad, Calgary
Hans Joachim Brown, Calgary
Mona J. Bryan, Edmonton
Leslie K. Brygadyr, Edmonton
Mary E. Buckley, Calgary
Margaret L. Burch, Edmonton
Tom P. Byrne, Edmonton
Helen M. J. Caklos, Blairmore
Nancy E. Carlyle, Blackfalds
Gordon R. Carnegie, Edmonton
John R. Caron, Edmonton
Hugh L.E. Chamberlain, Edmonton
Pauline R. Chevalier, Edmonton
Annie Marion Chop, Dixonville
Judy M. Cissell, Ponoka
Carol A. Collier, Edmonton
Ulrike, E. Conradi, Edmonton
Kenneth A. Cooper, Red Deer
Robert I. Coote, Edmonton
Brian Gray Crummy, Edmonton
Colleen L. Cummins, Winterburn
Barbara A. Curwen, Calgary
Colleen Linda Dagnall, Calgary
Barry J. Deeprose, Calgary
Kenneth C. Dewar, Edmonton
Marian J. Dey, Ardrossan
Loreen E. Doell, Edmonton
Elizabeth J. Dickson, Calgary
Brian W. Dippie, Edmonton
Cheryl S. Dow, Lethbridge
Morris B. Drapper, Edmonton
Kennth R. Dubeta, Edmonton
Donna L. Duchak, Calgary
Maurice I Dumont, Bonnyville
Robert B. Dunbar, Calgary
Mavis I. Dunford, Calgary
Eleanor G. Dupilka, Athabasca

Randy P. Dyck, Balzac
Christine E. Eder, Edmonton
Michael W. Edwards, Edmonton
Jennifer A. Ehly, Edmonton
Brian S. Ekstrom, Balzac
Dietor Ensslen, Edmonton
Barbara J. Erickson, Eckville
Sharon L. Ericson, Wetaskiwin
Gordon J. Esplin, Lacombe
Colin M. Evans, Cowley
Lorne L. N. Everndem, Lethbridge
Robert J. Fairbairn, Vegreville
Norman R. Fawcett, Bluffton
Maureen G. Fiddes, Gleichen
Patricia E. Finley, Stirling
Margaret, J. Fisher, Edmonton
Gordon J. Fitzpatrick, Drumheller
Leona A. Fix, Stettler
Russell B. Flewwelling, Hinton
Violet Fodor, Warburg
Robert A. Folinsbee, Edmonton
Donna G. Forbes, Edmonton
John D. Ford, Edmonton
Edgar F. M. Forster, Lethbridge
Robert G. Foulon, Camrose
Geoffrey G. Frank, Edmonton
Claire M. R. Fraser, Calgary
Joseph P. Freedman, Edmonton
Mary Friesenhan, Edmonton
Charles E. Frost, Edmonton
David A. W. Fustukian, Edmonton
Patricia E. Gander, Edmonton
James R. G. Gardiner, Edmonton
John William Gee, Edmonton
Evelyn S. Gerlach, Stettler
Dianne E. Gerun, Perryvale
Alexander I. Gibson, Westlock
Beverly M. Gietz, Pincher Creek
Brian A. Glanfield, Hoadley
Erik O. Goble, Waterton Park
Norbert Goebel, Ft. Saskatchewan
Myron Goldberg, Edmonton
Marilyn R. Goodman, Red Deer
John C. E. Greene, Edmonton
Marilyn M. Gregorwich, Westlock
Judith B. Gunning, Edmonton
Richard Allan Gusella, Calgary
Gail L. Haden, Calgary
Frederick W. Haeseker, Calgary
Marcia Ann Hagen, Clandonald
Donna M. Hamar, Lac La Biche
Dieter G. Hanlich, Jasper Place
Glen H. Harper, Edmonton
Jacqueline V. Harvie, Calgary
Robert M. Heise, Edmonton
Shirley I. Hennig, Edmonton
Harry G. Hermann, Hardieville
Patricia A. Himmelman, Calgary
Annalee M. Hirsch, Bawlf
Gordon Holt, Lethbridge
Douglas P. Horning, Edmonton
Margaret L. Huckvale, Edmonton

Erwin Huebner, Edmonton
Bonnie E. Hughes, Ponoka
Elaine D. Hughes, High River
Gary Horlick, Calgary
Kenneth H. Hutchinson, Edmonton
Frank B. Jacobson, Red Deer
Gary P. James, Edmonton
Thomas A. Jenkyns, Calgary
Dennis W. Jirsch, Edmonton
Mary E. Johns, Edmonton
Cheryl A. Jones, Dewberry
Ralph G. Jorstad, Taber
Inge-Edith Junkuhn, Edmonton
Judith Lynne Kales, Edmonton
Marjorie Edith Keller, Calgary
Beverly A. Kellett, Edmonton
Diane E. Key, Beverly
Darlene Mona King, Three Hills
Eugene Kizior, St. Francis
Lorne A. Klippert, Calgary
Virginia L. Knuston, Pincher Creek
Anne Lucille Koenig, Medicine Hat
Stanley M. Kohler, Calgary
Zoly J. Koles, Edmonton
Eileen O. M. Kosior, Black Diamond
Vera M. Kotylak, Edmonton
Egbert H. Krikke, Lacombe
Val J. Kulak, Edmonton
Gordon G. Kurio, Lethbridge
Lucienne G. Lambert, Guy
Thomas A. Landsman, Duffield
Arthur N. Larson, Medicine Hat
William Laskowski, Edmonton
Michael E. Laub, Calgary
Sheena E. Laycraft, Calgary
Ronald D. Lee, Edmonton
Donald G. Leinweber, Calgary
Carole E. A. Lencucha, Blairmore
Philip G. Lister, Edmonotn
John E. E. Lofkrantz, Cold Lake
Mary L. Losie, Medicine Hat
Robert H. Louie, Calgary
Alastair R. Lucas, Edmonton
Margot C. Lukas, Coaldale
Stuart R. Lumas, Edmonton
Derek J. McCune, Ft. Vermilion
Flora M. MacDonald, Calgary
√ Rene A. McElroy, Edmonton
Douglas W. MacFarlane, Coleman
Grace A. McGillivray, Provost
Antony R. MacGinnis, Calgary
Douglas G. MacKenzie, Calgary
Allister J. McPherson, Calgary
Allan D. MacTavish, Edmonton
Brenda M. Mallen, Edmonton
Sheila Malm, Vauxhall
Sylvia Malm, Vauxhall
Gerald F. Manning, Edmonton
Fay Eileen Martin, Camrose
Henry B. Martin, Edmonton
Denis P. Matisz, Lethbridge
Anne C. Matthews, Calgary
Calvin L. Merkley, Edmonton
Robert Edward Miller, Edmonton
Violet A. Miller, Edmonton
John F. Milner, Edmonton
Katherine H. Mitchell, Calgary
Kathleen J. Moon, Edmonton
Robert G. Moore, Seebe
Clifford G. Morgan, Edmonton
James M. Moscovich, Lethbridge
John D. Mulholland, Edmonton
Joan I. Neilson, Onoway
Kathleen E. Nichols, Edmonton
Ronald E. Nieman, Calgary
Phyllis M. Nimchuk, Edmonton
Joseph V. I. Noel, St. Lina
Garry W. Nyrose, Calgary
Frederick D. Parker, Calgary
Dilys M. Parry, Edmonton
Diane Helen Parsons, Westlock
Glenda B. Patterson, Tofield
Elsa L. Pearson, Bon Accord
Vera A. Pederson, Hughenden
Gerald R. Penner, Calgary
Dianne M. Pinder, Ft. MacLeod
Brian C. Plain, Calgary
Elizabeth Poss, Thorhild
Phylis M. Powell, Sexsmith
Claire M. Poulin, Edmonton
Gerald D. Powlik, Thorsby
John T. Price, Calgary
Peter J. Puchyr, Edmonton
Russell M. Purdy, Lethbridge
Ilona A. Raycheba, Edmonton
Audrey F. Rea, Edmonton
David J. Reece, Edmonton
Judith, E. Reeves, Edmonton
Henry D. Rempel, Edmonton
Paul F. W. Rutherford, Edmonton
Gwyneth M. Roberts, Calgary
Kerri D. Robertson, Edmonton
Donna M. Robinson, Edmonton
Scott M. Robson, Bawlf
Constance W. Roper, Hay Lakes
Earl L. Rosser, Silverwood
Donna Lynn Roxburgh, Edmonton
A. Ruth Runions, Edmonton
Paulette, C. E. Rypien, Coleman
William M. Saunders, Calgary
Lavera Mae Sawatzky, Westlock
Mervin R. Schafer, Olds
Sandra E. Scheierman, Vulcan
John H. Schmidt, Olds
Briane P. Schow, Cardston
Gilbert A. Schultz, Hay Lakes
Helmut Schwachhofer, Calgary
Dorothea H. Schwanke, Holden
Suzanne J. Semeniuk, Boyle
Daniel D. Sereda, Edmonton
M. Elaine Sereda, Edmonton
Alan E. Sharpe, Morrin
Eric H. Shelton, Edmonton
James K. Side, Edmonton
Edward E. Simbalist, Edmonton
Georgia J. Sinclair, Red Deer
John W. D. Sinclair, Edmonton
Eunice M. Sitler, Camrose
Sheila N. Smeltzer, Edmonton
Elisabeth Smit, Edmonton
Vivian P. Smith, Olds
Virginia K. Sorenson, Bawlf

Steven J. Sprysak, Edmonton
John W. Stamm, Edmonton
Katherine M. Stenger, Warburg
Deanne M. Stepchuk, Edmonton
Donald E. Stephens, Red Deer
Edith J. Stilwell, Calgary
Bonnie C. Strader, Calgary
Brian D. Sykes, Edmonton
Jerry A. Takacs, Edmonton
Robert W. Taylor, Magrath
Morris F. Tchir, Edmonton
Roland E. Theroux, Bonnyville
Margaret J. Thompson, Edmonton
Linda E. Thorssen, Calgary
Aubrey J. Tingle, Edmonton
Dorothy R. Tovell, Vermilion
Richard L. Treleaven, Lacombe
David L. J. Tyrrell, Duffield
Rosina Urch, Calgary
John Van Dyke, Edmonton
Job W. Vander Wekken, Lacombe
Hendricus J. Van Reede, Edmonton

Mark R. Venier, Coleman
Davis Walden, Edmonton
Douglas C. Walker, Edmonton
William P. Ward, Edmonton
Nola D. Watts, Lloydminister
Hans Wehrfritz, Edmonton
Kenneth Welsh, Milk River
Sharon Ruth Widlake, Galahad
Walter K. Wilkinson, Olds
Allan H. Wilson, Carstairs
Ronald A. Winter, Edmonton
Carol M. Wood, Banff
John Norman Wood, Calgary
Gordon L. Woodman, Westlock
Ronald N. Wright, Calgary
Dorothy Joyce M. Yarema, Holden
Gloria J. Yarmoloy, Vegreville
Malcolm J. Young, Edmonton
Vernon A. Zelmer, Warburg
Marie E. Zenko, Edmonton
John C. Zubis, Edmonton

AWARDS MADE BY OTHER INSTITUTIONS

The following awards are open in one Field of Study only:

ENGINEERING

Canadian Good Roads Association Scholarship
Helmut Keith Walker, Edmonton

NURSING

University of Alberta Hospital School of Nursing Alumnae Association Scholarship
Margaret MacDonald, Edmonton

SOCIAL WORK

Alberta Chapter, Canadian Foundation for Poliomyelitis and Rehabilitation Bursary in Social Work
Ian Walker, Vancouver, British Columbia

Imperial Order Daughters of the Empire Bursary in Social Work
Ian Walker, Vancouver, British Columbia

The Kinette Bursary in Social Work
James H. Allison, Edmonton

The P.E.O. Memorial Scholarship in Social Work
Lois Griffith, Rockyford

WILDLIFE MANAGEMENT

C.I.L. Wildlife Management Fellowship
Keith Reginald D. Mundy, Calgary

The above list of awards made by other institutions is not necessarily complete.

The list of candidates for degrees and diplomas appearing herein is subject to such corrections, with respect to both deletion and additions, as may be necessary.

Graduands whose names are preceded by an asterisk (*) are graduating in absentia.

Degrees

Admitted to the Degree of Doctor of Laws, Honoris Causa
√ Ian Nicholson McKinnon
√ The Honorable Chief Justice Colin Campbell McLaurin

Admitted to the Degree of Doctor of Philosophy
Candidates presented by Professor A. G. McCalla, Dean of the Faculty of Graduate Studies

Alan Fergus Brown, B.A., B.Ped. (Manitoba), M.Ed. (Alberta) "The Differential Effect of Stress-Inducing Supervision on Classroom Teaching Behavior." (Educational Administration), Calgary

Johan Frederick Dormaar, B.S.A., M.S.A. (Ontario Agricultural College) "A Study Concerning the Determination of Organic Phosphorus in Soils." (Soil Chemistry), Edmonton

*Jean-Yves, Drolet, B.Ed. M.Ed. (Laval) "A Study of the Impact of Demographic and Socio-Economic Factors on School Attendance Rates in the Province of Quebec from 1901 to 1959." (Educational Administration), Quebec

Frederick Enns, B.Ed., M.Ed. (Alberta) "The Legal Status of the Canadian Public School Board." (Educational Administration), Edmonton

Alwyn Bradley Ewen, B.A., M.A. (Saskatchewan) "Studies on Neurosecretion in the Alfalfa Plant Bug, *Adelphocoris lineolautus* (Goeze) (Hemiptera: Miridae)." (Entomology), Saskatoon, Saskatchewan

Graham Hugh Hunt. B.Sc. (Manitoba), M.Sc. (Alberta) "The Purcell Eruptive Rocks." (Geology), Edmonton

*Thomas Parry Jones, B.Sc. (North Wales) "Kinetic Studies on the Hydrolysis of Chromium (III) Complexes." (Chemistry), Wiltshire, England

Bernard Trueman Keeler, B.A. (Dalhousie), B. Ed. (Acadia), M.A. (Laval) "Dimensions of the Leader Behavior of Principals, Staff Morale and Productivity." (Educational Administration), Edmonton

Gerald Douglas Lutwick, B.Sc., M.Sc. (Dalhousie) "Column Efficiency in Gas-Liquid Chromatography." (Analytical Chemistry), Sarnia, Ontario

*John Melnyk, B.S.A., M.Sc. (Manitoba) "The Hybrids of Aegilops × Secale." (Genetics), Winnipeg, Manitoba

Herbert Daniel Peters, Th.B. (Canadian Mennonite Bible College), M.A. (Arizona) "An Experimental Application of the Concepts of Image and Plans to the Counseling Setting." (Educational Psychology), Saskatoon, Saskatchewan

Robert Arthur Ritter, B.E., M.Sc. (Saskatchewan) "A Theory of Thixotropic Behaviour and its Application to Pembina Crude Oil." (Chemical Engineering), Edmonton

*Mohammad Shamsuddin, B.Sc. (Calcutta), M.Sc. (Patna) "A Study of the Behaviour of Larval Tabanids (Diptera: Tabanidae) in Relation to Light, Moisture and Temperature." (Entomolgy), Bihar, India

Admitted to the Degree of Doctor of Education
*Laurence Maxwell Ready, B.A., B.Ed. (Saskatchewan) "The Preparation Needs of Superintendents in Large Administrative Units in Saskatchewan." (Educational Administration), Ottawa, Ontario

Admitted to the Degree of Master of Arts
*John Arbuckle, M.A. (Glasgow) "Phonology of the Volhynian German Dialect of The Edmonton Area." (General Linguistics), Ithaca, New York, N.Y.

Donald Asa Bancroft, B.Sc., Agric. (Alberta) "Equalization of Assessments in Alberta; The Principles, Authorities and Methods." (Economics), Edmonton

William Gerald Brese, B.A. (Alberta) "An Analysis of the Sulphur Industry in Alberta." (Economics), Edmonton

*Gerald Edward Brice, B.A. (Alberta) "The Development of Greek and Latin Elegiac Poetry and its Influence upon John Milton's Latin Elegies, Together with a Translation of these Elegies into English Verse." (Latin), Edmonon

Dorothy Jean Crouse, B.A. (Acadia) "A Study of Overinclusion and Underinclusion on the Picture Completion Test of the Wechsler Adult Intelligence Scale." (Psychology), Ponoka

David Gomer Fish, B.A. (Alberta) "Some Sociological Aspects of a Fluoridation Plebiscite." (Sociology), Edmonton

George Porges, B.A. (London), B.A. (Sir George Williams), B.Ed. (Alberta) "The Territorial Settlement of the Austrian Crown Lands, with Special Reference to Czechoslovakia, Italy and Yugoslavia, 1918-1919." (History), Edmonton

*George Samuel, B.A. (Alberta) "The Christ Figure in Blake and Shelley." (English), Edmonton

*Peter David Seary, B.A. (Newfoundland) "The Theme of Creativity in the Novels of Joyce Cary." (English), St. John's, Newfoundland

*Bruce William Wilkinson, B.Com. (Saskatchewan) "The Acceleration Principle in Economics Theory." (Economics), Calgary

*Albion Richard Wright, B.A., B.D.(Alberta) "Usury: Theory and Practice in the Medieval Period." (Economics), Drayton Valley

Admitted to the Degree of Master of Science

/ *Richard Alwyn Bramley-Moore, B.Sc., Mining Engineering (Alberta) "Oxidation of Stibnite." (Metallurgical Engineering), Thetford Mines, Quebec

*Syed Amir Bukhari, B.Sc. (Government College Lahore) "A Linear Programming Approach to the Interpretation of Earth Resistivity Data." (Physics), Edmonton

*John Robert Channon, B.Sc., Civil Engineering (Alberta) 'The Ultimate Load Capacity of Steel Frames." (Civil Engineering), Ponoka

*John Ivor Clark, B.Sc. (Acadia), B.Eng. (Nova Scotia Technical College) "A Study of Shrinkage Soil-Cement." (Civil Engineering), Granby, Quebec.

Roger Malcolm Evans, B.Sc., Arts (Alberta) "Courtship and Mating Behavior of Sharp-tailed Grouse (*Pedioecetes phasianellus jamesi*, Lincoln)." (Zoology), Madison, Wisconsin, U.S.A.

*James Woolley Gibb, B.Sc., Pharm. (Alberta) "The Effect of Certain Hormones on Tissue Respiration in the Rat." (Pharmacy), Ann Arbor, Michigan, U.S.A.

*William Anthony Gibbons, B.Sc. (Edinburgh) "Mercury Photosensitized Reaction of Cyclopentene." (Chemistry), Scotland

*Allan Angus Grunder, B.S.A. (Ontario Agricultural College) "Relationship Between Blood Group Genes and Economic Traits of Poultry." (Poultry Genetics), Glammis, Ontario

*Asif Uddin Hassan, B.Sc. (Delhi), M.Sc. (Panjab) "Some Experiments on Superfluidity in Liquid Helium II." (Physics), West Pakistan

*Yin Yew Hui, B.Sc. (Hong Kong) "Moment Generating Functional Equations of Certain Stochastic Learning Models." (Statistics), Hong Kong

*Sami Ahmad Sobhi Ibrahim, B.Sc. (Ain Shams University) "An Assessment of Wind Erosion Damage to Alberta Soils." (Soil Conservation), Davis, California

*Norman Clark Jamieson, B.Sc. (Edinburgh) "Preparation of Some Mercapto Monosaccharides." (Organic Chemistry), Edinburgh, Scotland

Robert Walter Jenkins, B.Sc. (British Columbia) "Meteorological Effects on Cosmic Radiation at Intermediate Depths Underground." (Physics), Calgary

Robert Kingsley Jull, B.Sc., Arts (Alberta) "Silurian Halysitidae of Western Canada." (Geology), Edmonton

Miles Steve Kuryvial, B.Sc, Agric. (Alberta) "Energy Digestibility Nitrogen Retention, Efficiency of Feed Utilization and Carcass Characteritics of Pigs Fed Varying Levels of Fat and Protein." (Animal Nutrition) Cranford

*Wilbert Emil Lentz, B.Sc., Agric. (Alberta) "Genetic and Environmental Variance and Covariance in Beef Cattle Performance Characteristics." (Animal Breeding), Edmonton

Mary Winifred McKay, B.Sc., Arts, (Alberta) "Mississippian Foraminifera of the Southern Canadian Rockies, Alberta." (Geology), Edmonton

*Ronald Ralph Marquardt, B.S.A. (Saskatchewan) "The Effects of the Water Soluble Components of Forages in the *in vitro* Cellulolytic Activity of Rumen Microorganisms." (Animal Nutrition), Edmonton

*Richard Graham Miller, B.Sc., Arts (Alberta) "A Coincidence Gamma-Ray Spectrometer with Applications." (Physics), Edmonton

Ellenor Ingrid Neumann, B.Sc., Arts, (Alberta) "Investigations of Reactions of Possible Analytical Value for Uric Acid." (Biochemistry), Edmonton

*Adery Catherine Alison Patton, B.Sc. (McGill) "The Approximate Distribution of the Ratio of the Square Successive Difference to the Square Difference in Observations from a Stationary Normal Markov Process." (Statistics), Edmonton

*Robert John Shivak, B.A., B.S.P. (Saskatchewan) "The Influence of Reserpine on Peripheral Vascular Responses to Hypercapnia." (Pharmacology), Ann Arbor, Michigan, U.S.A.

*Lakshmi Prakash Srivastava, B.Sc., Metallurgical Engineering (Banaras) "Martensite in Titanium, Zirconium and Titanium-Copper Alloys." (Metallurgical Engineering), Minneapolic, Minnesota, U.S.A.

Derek Bertram Swinson, B.Sc. (Queens, Belfast) "Variations in Extensive Air Showers." (Physics), Calgary

*Edward Lawrence Tomusiak, B.Sc. Engineering Physics (Alberta) "Two Electron Capture by Fast Alpha Particles in Helium." (Theoretical Physics), Montreal Quebec

*James Grant Tyner, B.Ed., Physical Education (Alberta) "Mechanisms Increasing Thyroxine Requirement of Rats in the Cold." (Physiology), Edmonton

Roy Yoshinobu Watanabe, B.Sc. (McMaster) "Geology of the Waugh Lake Metasedimentary Complex, Northeastern Alberta." (Geology), Edmonton

*Norman David Watkins, B Sc. (Leicester), M.Sc. (Birmingham) "Studies in Paleomagentism." (Geology), Menlo Park, California, U.S.A.

William Gordon Watt, B.Sc., Civil Engineering (Alberta) "Stabilization of a Highly Plastic Clay with Lime and Pozzolan." (Civil Engineering), Saskatoon, Saskatchewan

*Dorothy Lilian Weijer-Tolmie, B.Sc. (London) "Excretion Studies, Total Body Radiation, and Radiation to the Blood in Patients Treated with Radiocative Phosphorus (P^{23}) for Polycythemia Rubra Vera." (Clinical Radiation Biology), Sherwood Park

*Ronald Leslie Whitehouse, B.Sc., Dairying (Nottingham) "Disinfection with Sodium Hydroxide and its Application to Milk Production." (Dairy Microbiology), Ottawa, Ontario

*Celia Mary Yates. B.Sc., Pharm. (Glasgow) "Effect of Reserpine on the Reactivity of Isolated Rabbit Blood Vessels." (Pharmacology), Dumfries, Scotland

Admitted to the Degree of Master of Education

Raymond Clarence Carran, B.Ed. (Alberta) "A Study of Student and Adult Attitudes Towards the Technical Electives Program in Edmonton Composite High Schools." (Educational Administration), Edmonton

Milton Reinhold Fenske, B.Ed. (Alberta) "An Analysis of the Work-Week for a Sample of Central Alberta High School Teachers." (Educational Administration), Trochu

*Mary Bernardette Glennon, B.A. (Queens, Belfast), B.Ed. (Alberta) "An Investigation of the Relationships Between Two Speeded Tests of Visual-Motor Skills and a Measure of Reading Achievement." (Educational Psychology), Saskatoon, Saskatchewan

John Albert Jenkinson, B.Sc. (Liverpool), B.Ed. (Alberta) "The Educational Ideas of St. Augustine." (Educational Foundations), Edmonton

*Billie Eleanor Jean McBride, B.Ed. (Alberta) "The Parental Identifications of Adolescents." (Educational Psychology), Edmonton

George Adolph Mann, B.Ed. (Alberta) "Alberta Normal Schools: A Descriptive Study of their Development, 1905 to 1945." (Educational Foundations), Lethbridge

Neil MacLean Purvis, B.Sc., Arts (Alberta), "A Survey of Second Language Programs for English-Speaking Children in Grades One through Nine in Canadian Schools." (Secondary Education—Foreign Languages), Edmonton

*Steve William Radomsky, B.Sc., Arts (Alberta) "A Comparative Study of the High School Physical Science Programs for Two School Years, 1935-36 and 1959-60." (Secondary Education—Science), Edmonton

*Robert Henry Routledge, B.Ed. (Alberta) "A Study to Establish Norms, for Edmonton Public Secondary School Boys, of the Youth Fitness Tests of the American Association for Health, Physical Education, and Recreation." (Secondary Education—Physical Education), Edmonton

Joseph Francis Swan, B.A. (Queens, Kingston), B.Ed. (Alberta) "A Historical Survey of the Board of Reference in Alberta." (Educational Administration), Edmonton

*Ross Eugene Traub, B.Ed. (Alberta) "Social Desirability in the Rural High School." (Educational Psychology), Edmonton

Admitted to the Degree of Bachelor of Divinity

Candidates presented by Dr. E. J. Thompson,
Principal of St. Stephen's College

Stuart Ledlie Munro, B.Sc., Lac La Biche
*Keith McKay Page, B.A., White Sulphur Springs, New York
William Andrew Sayers, B.A., Calgary

Admitted to the Degree of Doctor of Medicine

Candidates presented by Professor J. S. Thompson,
Assistant Dean of the Faculty of Medicine

Joseph Bertrum Busheikin, Edmonton
George Leon Abel Douchet, Edmonton

Admitted to the Degree of Doctor of Dental Surgery

*Earle Montcalm Bremner, Edmonton
*David Allan Wolfe, Edmonton

Admitted to the Degree of Bachelor of Arts

Candidates presented by Professor D. E. Smith,
Dean of the Faculty of Arts and Science

David James Anderson, Calgary
*Peter Edward Arabchuk, Edmonton
*Arnold Bernstein, Edmonton
*Andre Brock, Edmonton
*Bonnie Noreen Bryans, Edmonton
Robert MacMillan Cairns, Calgary
Edward Ronald Rumney Carruthers, Calgary
*Daniel Chan, Red Deer
*Donald Charles Clayton, Edmonton
*Sister Gabriel, Edmonton
 (Eveleen Collins)
Phyllis Edith Mary Darby, Edmonton
Norma Lorraine Davidson, Edmonton
*Barbara Elizabeth Downs, Edmonton
*Joseph Doz, Edmonton
Anthea Lynne Murray Dykes, Wetaskiwin
Mervyn Norman Guy Eastman, Edmonton
*Lawrence Ewanchuk, Hairy Hill
James Lambert Foster, Red Deer
Margaret Elizabeth Dianne Gussow, Edmonton
Jean Marie Haddow, Edmonton
Adolf Kurt Heise, Redcliff
*John Matthew Hrynew, St. Albert
Inger Boesen Jacobsen, Edmonton
Joyce Isabelle James, Edmonton
*Alan Boyd Johnson, Edmonton
*Arthur Otto Jorgensen, Bonnyville
James Arthur Kelly, Edmonton
Harry Kleparchuk, Edmonton
Stephanie Anne Lawrence, Red Deer
*Lynette Verna Lesnik, Fort St. John, B.C.
*Katherine Stuart McAllister, Edmonton
Grace Elaine McEachern, Edmonton
James Kent MacKinlay, Edmonton
Robert William Miller, Coutts
Janet Mary Morrison, Edmonton
*Douglas Leslie Oke, Edmonton
*Dennis Albert Person, Edmonton
Arta Blanche Pilling, Calgary
Margery Helen Ramsay, Calgary
*Robert Hart Robinson, Camrose
Leo Cameron Ross, St. Albert
*Louis Paul Salley, Edmonton
*Arthur Otto Stinner, Edmonton
*Cecil Archibald Woodward, Edmonton
*Frederick Garth Worthington, Edmonton
*David James Wright, Edmonton
Austin Wilbur Youngberg, Edmonton

Admitted to the Degree of Bachelor of Science in Arts and Science

William Sanderson Adams, Edmonton
Patricia Constance Adshead, Edmonton
David William Antoniuk, Edmonton
Harold Robert David Beckman, Edmonton
George Archibald Campbell, Edmonton
Michael Joseph Charuk, Myrnam
*Willem Dammeyer, Edmonton
*Agnes Marie Dick, Edmonton
*Edward Wayne Dutton, Redcliff
Brian Robert Farrell, Edmonton
Maria Flak, Edmonton
Florence Mary Anne Hancock, Gibbons
*Gerald Gladwyn Harrington, Spirit River
*Herbert Edward Herunter, Vancouver, B.C.

*Darrell Clinton Hockett, Edmonton
*Lal Harry Jaglalsingh, Edmonton
*Michael Douglas Macdonald, Victoria, B.C.
Roger Allen McNabb, Medicine Hat
*Elizabeth Dianne McNaughton, Edmonton
*Paul L. Manley, Edmonton
*Turid Liesel Minsos, Edmonton
James Campbell Monro, Calgary
*Myrna Eileen Nichols, High River
Gordon Andrew Noel, Lethbridge
*Richard Dennis Odland, Calgary
*Ralph Henry Ohrn, Edmonton
*Rudolf Peters, Olds
Glen Harmon Pierson, Taber
Stephen Russel Ramsankar, Edmonton
Father Josef Franz Paul Schmelz, Calgary
Andrew Ray Schmidt, Calgary
John William Slemko, Calgary
Christopher James Sparks, Calgary
*John Jacob Sribney, Calgary
Bruce Gavin Stewart, Edmonton
William George Tatton, Calgary
Gerald Mervyn Tobert, Edmonton
Morris Ralph Treasure, Edmonton
Gerald Linwood Weitzel, Calgary
Roger Onus York, Edmonton

Admitted to the Degree of Bachelor of Music

*Isobel Patricia Clowes, Danville, Quebec

Admitted to the Degree of Bachelor of Science in Civil Engineering

Candidates presented by Professor G. W. Govier, Dean of the Faculty of Engineering

Bruce Imrie Bryson, Edmonton
Archibald Fredrick Dick, Edmonton
James Gordon Elliott, Edmonton
Howard Douglas Hepburn, Edmonton
*Laurie Yook Bow Lee, Calgary
*Hugh John Stollery, Armena
*Gordon Alexander Wallace, Alsask, Saskatchewan

Admitted to the Degree of Bachelor of Science in Electrical Engineering

*William Borusiewich, Ottawa, Ontario
*Kwok-Yim Lau, Edmonton
*Paul Leo Lingas, Edmonton
*Harmon David Mah, Bashaw
Terry Lyal Tomlinson, Calgary

Admitted to the Degree of Bachelor of Science in Engineering Physics

*Paul Henry Karvellas, Edmonton
Kenneth Ronald Shultz, Edmonton
Walter Peter Sturm, Edmonton

Admitted to the Degree of Bachelor of Science in Mechanical Engineering

Taichi Jack Fujino, Coaldale
*Dennis Arden Jensen, Edmonton

Admitted to the Degree of Bachelor of Science in Metallurgical Engineering

*Howard Siebert, Lister, British Columbia

Admitted to the Degree of Bachelor of Science in Petroleum Engineering

Donald Harvey Bickell, Edmonton
Edward Percy Webb, Drayton Valley
Gordon Robert Scott, Calgary
*James Rush, Peers

Admitted to the Degree of Bachelor of Science in Medical Laboratory Science

Candidates presented by Professor J. S. Thompson, Assistant Dean of the Faculty of Medicine

Kathleen Irene Rusnak, Edmonton
Margaret Patricia Shand, Edmonton
Geraldine Georgia Wilson, Edmonton

Admitted to the Degree of Bachelor of Science in Agriculture

Candidates presented by Professor C. F. Bentley, Dean of the Faculty of Agriculture

*John Robert Andrew, Ardrossan
Arthur Dahlton Whitehair, Olds

Admitted to the Degree of Bachelor of Education

Candidates presented by Professor H. T. Coutts, Dean of the Faculty of Education

With First Class General Standing
John Charles Macdonell, Cochrane
Emerson Ross Shantz, Didsbury

Shawne Allen, Calgary
Francis Garfield Anderson, Calgary
Kay Anderson, Huxley
*Ronald Hjalmar Anderson, Edmonton
Kenneth Powell Armitage, Edmonton
Mary Wynne Ashford, Vancouver, B.C.
Ruth Alice Auburn, Provost
Joyanne Kae Baker, Edmonton
Eugene Elmer Balay, Bow Island
William Stanley Baranyk, Olds
*Frederick Rupert Barber, Camrose
*George Langdon Bayliss, Calgary
Charles Ross Beggs, Edmonton
Peter Belma, Edmonton
*Axel Loren Benzon, Calgary
Irvin Milton Besler, Calgary
Carl Hervey Blumer, Edmonton
*Maurice Edmond Bourgoin, Edmonton
Hannah Genevieve Bradley, Calgary
*Edmund Gordon Breitkreutz, Stony Plain
Pearl Emma Brooks, Warburg
Doris Elizabeth Brosseau, Three Hills
Helena Myrtle Brown, Didsbury
Stephen Jeremiah Buckley, Calgary
Lorne Wilfred Bunyan, Beisecker
*Robert Findlay Burch, Edmonton
William Henry Burch, Innisfail
*Alice Joyce Cade, Lougheed
Mary Anne Theresa Calon, Michichi
Eleanor Campbell, Provost
Glen Norman Carmichael, Spruce Grove
*Sister Sheila Mary Cassidy, Edmonton
*John James Niddrie Chalmers, Edmonton
John Alexander Charnetski, Lethbridge
Jean Ann Chmilar, Berwyn
*Andrew Kinloch Clark, Calgary
Charles David Clark, Clive
Matthew Melvin Coates, Innisfail
Peter Victor Coldham, Calgary
Isabel Shellard Cole, Calgary
*Ambrose William Comchi, Winnipeg, Manitoba
*Peter Gilbert Connellan, Innisfail
*Sister Clare Marie, Allan, Saskatchewan
(Mercedes Conroy)
Eugene William Corry, Hairy Hill
James Nicholas Courtney, Duffield
Clifford Alexander Cummins, Bow Island
Gaston Philibert Curial, Edmonton
Larry Evans Dahl, Lethbridge
Gerald Ernest Dahms, Ponoka
Eleanor Ione Damkar, Standard
*Edwin Robert Daniels, High Prairie
*Sister Florence Marie Marguerite Dansereau, Edmonton
Kay Davis, Grande Prairie
Elvin Cecil Dayman, Calgary
Geraldine Margaret DeMaere, Granum
Alton Edward Dennis, Camrose
John Daniel Derrick, Calgary
Fay Avis Dersch, Calgary

*Sister Mary Felicity, Edmonton
(Joan Edna Diederichs)
Russell Anthony Dolinski, Willingdon
Edward William Dowling Rimbey
*Fred Joseph Dumont, Kinuso
Jeannie Beverley Dykstra, Stavely
Elizabeth Ann Edwards, Edmonton
Alvin Ewald Effa, Calgary
Jane Theresa Eisler, Grande Prairie
*Robert John Enders, Stony Plain
*John Enns, High Prairie
Edwin Albert Ernst, Calgary
*Rosalie Hildegard Evanechko, Leduc
Steve S. Faminow, Calgary
*Frank Walter Featherstone, New Norway
*Sister Clementine Brigid Feist, Lethbridge
*Florence Betty Flach, Edmonton
Marshall Fodchuk, Willingdon
Bernard Thomas Fossen, Carbon
Ella Louise Freeman, Ponoka
*John Lorne Frey, High River
Joan Louise Freypons, Drumheller
Sister Maureena of Sion, Moose Jaw, Saskatchewan
(Patricia Gertrude Fritz)
Fredric Leo Gainer, Banff
Irene Mary George, Bowness
*William John Girard, Edmonton
*Sister Marie Florena de Sion, Winnipeg, Manitoba
(Mary Alice Godin)
*Father Guy Goyette, Fahler
Helen Smith Graham, Saskatoon, Saskatchewan
*Kazimer Joseph Gurski, Edmonton
*Autumn Lois Haggerty, Camrose
*Eleanor Blanche Hall, Bon Accord
Jack Otto Handel, Lloydminster
Muriel Mary Hansen, Edmonton
*Marvin Lee Harris, Red Deer
Edith Elizabeth Hart, Edmonton
Roy Edward Hartling, Edmonton
*Beverley Jean Harvie, Edmonton
Marcus Edward Heck, Claresholm
*Joyce Anne Hendrickson, Camrose
*Louis Hochachka, Clive
Edna Loretta Hough, Edmonton
Mary Hryciw, Grande Prairie
*Stephen Reginald Hrynewich, Bassano
Ichio Ibuki, Lethbridge
Florence Elizabeth Irwin, Edmonton
*Donna Marion Jackson, Edmonton
Mabel Janet James, Edmonton
*Hermann Wilhelm Janzen, Edmonton
*Henry Reginald Jeffers, Innisfail
*Dorothy Eliza Jones, Medicine Hat
*Richard Raymond Joy, Grande Prairie
Jenny Keegstra, Alhambra
*Eunice Irene Kenney, Edmonton
John Alexander Kerr, Edmonton
Margaret Joneta Kjorlien, Camrose
*Tony Korble, Edmonton
Steve Korchinsky, Edmonton
Anne Margaret Kornelsen, Coaldale

Michael Nick Kowalchuk, Edmonton
Alex Kozmak, Edmonton
Donald Glen Kross, Edmonton
*Father Bernard Kuefler, Edmonton
*Lawrence Kuly, Edmonton
*Soeur Madeleine-de-l'Espérance, Edmonton
  (Madeleine Lafond)
*Benjamin Alexander Laidlaw, Edmonton
Edna June Lapp, Eckville
Helen Elizabeth Lavkulich, Edmonton
Jack Flemming, Layton, Trochu
*Francis Whitfield Lee, Sheerness
*Sister Blanche Lemire, St. Albert
  (Blanche Florette Lemire)
James Ernest Lenz, Calgary
Inez Noreen Liebreich, Lomond
*Edmund Royall Long, Jasper
*Hugo Gerald Loran, Edmonton
*Kathleen Elizabeth Lougheed, Red Deer
Robert Dean Lund, Boyle
*Myrtle Lucille Lundy, New Sarepta
*Georgina Marie McCowan, Hay River N.W.T.
Olive Evangeline McCreary, Edmonton
*Arthur Duncan McCue, Sexsmith
Florence Mary Margaret Macdonald, Calgary
*Hugh Andrew Macdonald, Lloydminster, Saskatchewan
*James Oliver MacInnis, Edmonton
*Donald Arthur MacIver, Whitecourt
Edwin Louis McIvor, Edmonton
*Donald Angus MacKenzie, Edmonton
Robert Douglas McLean, Calgary
Norma Joan McLenahan, Calgary
Robert James McLuskey, Calgary
*Floyd Walter McMillan, Lac La Biche
*Robert Hector MacQuarrie, Edmonton
*Daisy Maduram, Edmonton
*Elias Kost Makowichuk, Lac La Biche
Nick Marchak, Edmonton
Maria-Rose Mathieu, Trochu
*Sister Marilyn Margaret Matz, Edmonton
James Meehan, Scapa
Ellenmie Ann Melcher, Rockyford
Henrietta Margaret Miller, Blackfalds
*James Leigh Miller, Acme
Eileen Olive Moisey, Edmonton
*Father John Ivan Molnar, St. Louis, Missouri
*Murray Frederick Moore, Brampton, Ontario
Mary Helen Morrison, Red Deer
*Terrance Roger Mott, Edmonton
Joseph Thomas Mould, Lethbridge
*Lawrence Edward Mutual, Edmonton
*Adelle Gail Muzyka, Edmonton
Verna Elizabeth Nagel, Edmonton
Abraham Nikkel, Calgary
*David Robert Owen Noel, Edmonton
Robert Edmond Norton, Calgary
*Denis O'Driscoll, Edmonton
Nathan John O'Hare, Ralston
*David Herbert Oke, Toronto, Ontario
*Father William Oruski, Medicine Hat

Gordon Leslie Oswald, Edmonton
Edward William Overbo, Kinsella
*Louis Ludwig Pade, Paddle Prairie
Ramdhanee Samuel Gouriesankar Pagee, Dewberry
*Valentine Joseph Pailer, Thorsby
Monika Ingeborg Pallat, Calgary
Irene Palmer, Calgary
Philip James Patsula, Edmonton
Glen Allan Patterson, Calgary
Harry P. Pawliuk, Edmonton
Mike Pawliuk, Two Hills
Gordon Leonard Peers, Edmonton
*Gabrielle Jeanne Pelchat, Medicine Hat
John Roy Pengelly, Delburne
Sister Loretta, Calgary
  (Marjorie Constance Mary Perkins)
*Harry Petryshen, Marwayne
Max Gordon Pharis, Nobleford
*Michael Porcsa, Edmonton
Gordon Wayne Harry Price, Calgary
Philip Arthur Quinney, Berwyn
Joseph Quintilio, Bellevue
*Solomon Wilfred Ramsankar, Edmonton
*Robert Douglas Ramsay, Manly, Australia
Alexander Ratsoy, Calgary
Charles Thomas Reaume, Forest Lawn
Napoleon Appolinary Rebryna, Edmonton
*John Paul Redd, Medicine Hat
Edwin Redecopp, Medicine Hat
*Anthony Francis Reghelini, Kitscoty
*Father Albert Frederick Reiner, Edmonton
Victor Rempel, Bowness
*Berna Jean Rhoades, North Star
*Marion Jane Elizabeth Richmond, Faust
William Roy Riley, Edmonton
Charles Edward Roberts, Edmonton
John David Roberts, Calgary
Anton Edward Rogalsky, Lamont
*Melvin George Rude, Edmonton
Eugene Sadoway, Hairy Hill
Lorraine Carol Salt, Edmonton
*Robert Paul Sanche, Millet
Norman Arnold Sande, Edmonton
Andrew John Schaufert, Lethbridge
Elisabeth Christine Schmidt, Ponoka
John Roland Schoepp, Edmonton
Margaret Seelye, Red Deer
Marlene Mary Shearer, Edmonton
*Sister Marie Ste. Christine, Edmonton
  (Lucy Evelyn Sheehan)
*Sister Francis Joseph, Edmonton
  (Anne Therese Sheridan)
Nicholas Peter Sidor, Edmonton
George Delbert Lea Simpson, Calgary
Jeryl Helane Sims, Veteran
Joseph Angelo Siqueira, Spirit River
Michael Boris Skapa, Edmonton
*Dorothy Ethel Smyth, Unity, Saskatchewan
Katherine Mae Sokoluk, Calgary
*Alice Sophie Sprado, Edmonton
James Dale Stafford, Ponoka

Joseph Aziel Stevenson, Edmonton
Ella Margaret Stewart, Edmonton
Rose Strembiski, Edmonton
Marguerite Laurette Strip, Calgary
Joseph Swan, Medicine Hat
*Steven Raymond Switlick, Wetaskiwin
John Jack Switzer, Edmonton
*Marshall Theodore Syska, Edmonton
Gabriel John Tajcnar, Stettler
Dorothy Fern Tetley, Red Deer
George Thiessen, Calgary
*George David Thomas, Edmonton
*Norah Emaline Thomas, Edmonton
*Dennis Edward Thompson, Calgary
Mary Thompson, Edmonton
*Kathleen Elizabeth Tomyn, Edmonton
Mike Harry Tomyn, Vegreville
Hope Tower, Arborfield, Saskatchewan
*Harry Douglas Trace, Edmonton
*Lawrence Andrew Truckey, Onoway
*Ernest Edward Turnbull, Edmonton
Irene Van, Medicine Hat
*Paul Leonard Voegtlin, Ryley
Edna Helena von Hollen, Red Deer
*Hans Ernest von Stackelberg, Ponoka
Warren David Waddell, Edmonton
*Violet Rose Walker, Saskatoon, Saskatchewan
*Robert Allan Wall, Peace River
Carol Anne Waston, Calgary
Elaine Agnes Whelihan, Edmonton
Marie Ella Whitehorne, Lethbridge
Harold John Wiebe, Calgary
Harvey Neil Wilkinson, Calgary
Thelma Jean Wilkinson, Calgary
Edith May Willie, Calgary
*Donald Everett Wilson, Edmonton
Anthony Valentine Winkler, Pincher Creek
*Adele Rosina Gulzow Wright, Fort Vermilion
*Cora Marlene Wutzke, Calgary
*Nicholas Zakordonski, Vegreville
Steve Mike Zuk, Edmonton

Admitted to the Degree of Bachelor of Education in Industrial Arts

Angus George Scarlett, Calgary

Admitted to the Degree of Bachelor of Science in Pharmacy

Candidates presented by Professor M. J. Huston,
Dean of the Faculty of Pharmacy

Lillian Joan Bielech, Derwent
*Donald Allan Butler, Edmonton
*Kathleen Feduniw, Hairy Hill
*Donald Stewart Jones, Port Arthur Ontario
*Jean Mah, Elk Point
Stewart Raymond Nickerson, Edmonton
Marshall Lawrence Serediak, Edmonton
Norman Anton Zacharuk, Banff

Admitted to the Degree of Bachelor of Commerce

Candidates presented by Professor H. Harries,
Dean of the Faculty of Commerce

Robert Stewart Ash, Calgary
Clifford Douglas Bristow, Spruce Grove
*Leo Isidor Superstein, Edmonton
*Ernest Anthony Tadla, Saskatoon, Saskatchewan

Admitted to the Degree of Bachelor of Science in Nursing

Candidates presented by Professor R. L. McClure,
Director of the School of Nursing

*Sandra Lillian Allan, Calgary
*Doreen Marjorie Gardiner, Calgary

Admitted to the Degree of Bachelor of Science in Household Economics

Candidates presented by Professor E. L. Empey,
Director of the School of Household Economics

Patricia Joan Adamson, Lacombe
Donna Anne Poohkay, Edmonton

Admitted to the Degree of Bachelor of Physical Education

Candidates presented by Professor M. L. Van Vliet,
Director of the School of Physical Education

Barbara Joan Heaps, Winnipeg, Manitoba
Wayne Clark Kotch, Edmonton
David Joseph Sande, Edmonton
*Theodore Bohdan Scherban, Edmonton
Irvin Benjamen Servold, Edmonton
*Gordon Severin, Edmonton

Diplomas

Diploma in Public Health Nursing
*Candidates presented by Professor R. L. McClure,
Director of School of Nursing*

*Veronica Victoria Anne Heinish, Edmonton

Diploma in Teaching and Supervision in Schools of Nursing
Anna Marie-Claire Laberge, St. Paul

The General Faculty Council of the University has also approved the award of diplomas to the following students who do not attend Convocation:

Diploma in Nursing

Elizabeth Achtemichuk, Myrnam
Emiko Adachi, Coalhurst
Leah Denise Baker, Edmonton
Carol Louise Bullock Bennett, Raymond
Eleanor Mary Betts, Calgary
June Anne Blackwell, Edmonton
Bette Joan Blackwell, Edmonton
Myrna Mae Brimacombe, Maidstone, Saskatchewan
Sandra Margaret Easton Brooks, Edmonton
Johann Hill Moreton Brown, Canmore
Harriet Frances Bruce, Edmonton
Vivian May Swischuk Bullen, Calgary
Helen Josephine Cameron, Edmonton
Margaret Anne Chamberlain, Regina, Saskatchewan
Darlene Isabel Chayer, Sunnynook
Anne Patricia Edith Clark, Yorkton, Saskatchewan
Sandra Jean Common, Edmonton
Lillian Coxon, Edmonton
Ansley Elizabeth Day, Edmonton
Sylvia Nancy Deputan, Poe
Jean Marie Donaldson, Medicine Hat
Margaret Helen Draper, Edmonton
Victoria Fedorak, Willingdon
Mutsuko Furuse, Coaldale
Violet Nadjah Fyk, Edmonton
Kathryn Elizabeth Jean Gaetz, Lethbridge
Barbara May Gammon, Grande Prairie
Eleanor Margaret Gaychuk, Edmonton
Florence Audrey Gilmour, Edmonton
Marjorie May Godsell, Edmonton
Phyllis Marie Hanson, Red Deer
Margaret Pearl Hawkins, Red Deer
Sonja Ellen Heney, Edmonton
Lorna Beth Hughes, Calgary
Linda Gail Hydorn, Edmonton
Laurene Isabel Jickling, Provost
Deanna Margaret Kendze, Condor
Katherine Deanna Kozak, Lac La Biche
Florence Ann Kozniuk, Delburne
Adele Ethel Kunigiskis, Thorhild
Bertha Kurylowich, Grimshaw
Georgie Clare Lundberg, Red Deer
Kathleen Audrey McCauley, Lloydminster
Ruth Letitia McCourt, Kitscoty
Stella Marie McKibbon, Ponoka
Gloria Elizabeth McLean, Edmonton
Patricia Ann Maher, Edmonton
Marjory Ruth Matthews, Irma
Juane Elsie Mattson, Bellevue
Elaine Helen Meighen, Edmonton
Georgia Gayle Millbank, North Battleford, Saskatchewan
Arlene Helen Muraca, Wetaskiwin
Eleanor Jean Newnham, Jarvie
Olive Lynn Nielsen, Streamstown
Cecelia Grace Nimchuk, Redwater
Sherry Carol Nolan, Tofield
Sonja Dorthea Nyback, Camrose
Madeleine Agnes Nysetvold, Chauvin
Vera Palas, Coaldale
Vilma Palas, Coaldale
Janet Diane Park, Edmonton
Hazel Winnifred Patrick, Lacombe
Lynda Dawn Patteson, Calgary
Carol Anne Petersen, Stettler
Phyllis Mabel Quon, High River
Diane Bernice Richard, Edmonton
Kay Arlene Ritchie, Lloydminster
Jane Elizabeth Ruschiensky, Markinch, Saskatchewan
Olive Annie Saliwonchuk, Reno
Marilyn Emma Schnell, Camrose
Perry Lynne Shannon, New Westminster, B.C.
Darlene Lenore Sills, Edmonton
Sylvia June Skoreyko, Smoky Lake
Mary Louise Skrove, Rolling Hills
Margaret Mae Smith, Lousana
Marjorie Mabel Soice, Lethbridge
Edith Marilyn Stewart, North Edmonton
Virginia Ruth Thompson, Edmonton
Elizabeth Bernice McElman Thornton, Edmonton
Stephanie Marguerita Marie Wacko, Jasper
Lucinda Maxine White, Devon
Norma Ruth Wolfe, Edmonton

Music

Associate Diploma in Music

Patricia Mae Colvin (Pianoforte-Performer, first-class honors), Calgary
Mary Nan Dutka (Pianoforte-Performer, honors), Canmore
Maria Wong (Pianoforte-Performer, pass), Sioux Falls, S.D., U.S.A.
Karen Olsen (Singing-Performer-Teacher, pass), Calgary

PROFESSIONAL CERTIFICATION

CERTIFICATES OF POSTGRADUATE QUALIFICATION IN DENTISTRY

Orthodontics

Dr. Van Evangelos Christou
Dr. Robert Matthew Perry

Dr. George Wesley Street

AGROLOGY

MAY, 1961
  Bruins, Gordon Gunster
  Gerdts, Walter-Hans J.
  Hanson, Morris
  Hawkins, Murray H.
  Kennedy, William F.

  McKenzie, R. Colin
  Peters, Daniel Irvine
  Ross, Gordon Alexander
  Stringam, Bryce Coleman
  Wroe, Robert Alistair

JUNE, 1961
  Guccione, Gioacchino, M.
  Morton, Kenneth, M.

  Robson, James Hunter
  Van Reekum, Gysbert

JULY, 1961
  Wilkes, Reginald T.

AUGUST, 1961
  Russell, Glenn C.

SEPTEMBER, 1961
  Grieve, Clarence Melross

OCTOBER, 1961
  Vigfusson, Norman V.

ARCHITECTURE

MAY, 1961
  Milan, John F.

JUNE, 1961
  Munzel, Alexander Oskar H.

SEPTEMBER, 1961
  Atkins, Gordon Lee

CHARTERED ACCOUNTANTS

JUNE, 1961
  Aagaard, Robert
  Ambrose, Henry E.
  Baird, Oliver Perry
  Berlanguet, George A.
  Bertsch, Harvey W.
  Blacker, Wililam J.
  Calloway, Fred
  Cameron, Ronald W.
  Cathro, David W.
  Caukill, Norman H.
  Chapman, Ronald A.
  Chivers, Harry D.
  Clearwater, Ronald
  Dalmer, Roger D.
  Denham, Ross A.
  Duda, Melvin L.
  Eurchuk, Eugene W.
  Evans, George B.
  Evans, James E.
  Eykelbosh, Thomas
  Faibish, David M.
  Frazer, James S.

  Gillespie, David L.
  Gordon, George W.
  Gorecki, Ernest
  Gourley, Alexander
  Graschuk, Harry
  Grossman, Frederick
  Halford, William
  Hamilton, Gerald
  Hartley, James
  Heron, David F.
  Holt, Charles M.
  Huff, Stewart, G.
  Jacobs, Robert J.
  Johnstone, Kenneth
  Jones, Thomas
  Jorgenson, Edgar
  Kay, Gary G.
  Kobie, Clayton
  Laslop, Harry W.
  Levine, Israel
  Matear, John Remi
  Maxwell, Hugh W.

McClelland, Robert
McGregor, Edward
McKague, Ronald
Palmer, Miles W.
Phillips, Harley D.
Riezebos, Norman
Robinson, Ronald G.
Rocque, Roland
Sidjak, Stanley S.
Strang, Vernon H.

Thompson, Kenneth W.
Thorpe, John B.
Trimble, Gary D.
Valentine, C. Peter
Walsh, Joseph N.
Watkin, Robert W.
Watson, Peter H.
Weitz, Perry
Willis, David Lea
Zingle, Del Franklin

DENTISTRY

JUNE, 1961
  Horton, Donald William
  Miller, Leo Douglas

Payne, Vernon Kent

JULY, 1961
  Newbern, Carl Hulie
  Southwood, Harold T.

Tueller, Vern M.

AUGUST, 1961
  Garner, Arthur D.
  Grabow, Leon Stanley

Siemens, Edwin Albert

SEPTEMBER, 1961
  McNee, Sydney John

ENGINEERING

MAY, 1961
  Burden, Hanford P.
  Cairns, Alvin W.
  Campbell, Neil
  Fox, Allan B.
  Gould, Charles D.
  Hasegawa, William N.
  Holmes, Thomas F.
  Hope, James R.
  Kelsay, William R.
  Kondi, Andy G.
  Morris, Lawrence R.

Patterson, Ralph W.
Porter, William A.
Ritenburg, Lorne H.
Ross, Ian S.
Seraphim, Andrew F.
Silgailis, Igor V.
Sigvaldason, Oskar, T.
Steed, Robert D.
Williamson, William R.
York, David

JUNE, 1961
  Aitken, Henry S. H.
  Beitz, Robert S.
  Buckingham, H. W.
  Burger, Robin J.
  Davidson, Russell M.
  Dzidrums, Martin
  Gass, Nicholas J.
  Hatch, Derek H.
  Hunter, John P.
  Koles, Stephen
  Kovacs, Stephen

Miles, Bruce M.
Morrison, Charles H.
Owen, John R.
Pope, Stephen H.
Quilty, Stanley M.
Richardson, Robert T.
Smith, George W.
Verner, Richard C.
Vornbrock, William J.
Walters, Roy M.
Yungblut, Glenn R.

JULY, 1961
  Belovich, Eugene M.
  Berndtsson, Norman G.
  Bolt, Leon H.
  Bookbinder, Leonard
  Bowsfield, Blaire L.
  Browne, Gary W.
  Chernoff, George M.
  Cox, Victor R.
  Davidson, Lawrence
  Elkington, Gerald W.

Ellis, Barrey A.
Ferrier, Donald R.
Forde, William E.
Fossenier, Germain B.
Gallon, Alan V.
Gemmell, Kenneth G.
Gray, Nigel G. D.
Gray, William J.
Greentree, William W.
Hanert, Alvin E.

Hindmarch, Kenneth J.
Hookings, Paul H. H.
Jepson, Derek, R.
Kovacs, Peter P.
Kraychy, Paul N.
Laatsch, Harold K.
Lipovschk, Milorad S.
Louie, Edward G. L.
Lukomskyj, Paul
McApline, Robert A.
McArthy, Rosscoe, G.
McMillan, Gordon W.
Maguire, Leonard J.
Malquist, Reijo T.
Maze, James R.
Metcalf, Wilbur B.
Middleton, Victor H.
Moore, John A.
Moriarity, Merton T.
Norquay, Ian P.
Oldham, William K.

Onyskko, Steve
Palmer, Gary H.
Parham, Kenneth R.
Parker, Gilbert M.
Patterson, Donald W.
Perry, Brian R.
Peterson, Andrew W.
Pinkney, Robert B.
Reichert, Joseph F.
Schneider, Louis J.
Schuett, George H.
Storme, Allan R.
Tapuska, William A.
Thrun, Eric R.
Turner, Richard L.
Twa, Craighton O.
Van der Linden, Jerry R.
Wahl, Herman E.
Walker, Robert
Wichert, Edward

AUGUST, 1961

Bamford, Michael, A.
Blanchette, Raymond L.
Bohen, Arthur F.
Cornish, David W.
Cowper, Robert B.
Croome, Norman C.
De Pan, Raymond T.
Fodor, Emery
Gallagher, John J.
Gribben, Hugh J.
Hamilton, Alexander B.
Hipp, Peter

Hnatiuk, Ernie W.
Hornford, Herbert E.
Link, Theodore W.
McEwen, Philip K.
Maunder, James F.
Menard, Harvey
Platt, Ronald L.
Porochnuk, William N.
Porter, William R.
Shaflik, Rudolph P.
Steeves, Kenneth L.

SEPTEMBER, 1961

Anderson, Lauchlin A. R.
Barlow, Charles B. Jr.
Barry, Charles, L. E.
Bierlmeier, William G.
Borwick, Ian R. H.
Burrell, John K.
Butterfield, Floyd N.
Carle, Donald W.
Carlisle, Claude
Charleston, James
Dawson, John G.
Denisik, Russell W.
Drucker, Emil
Farrell, Douglas D.
Fogarasi, Marie
Frankfurt, William W.
Halls, Charles R. G.
Horn, Thomas R.
Huisman, Jan
Humphreys, Reginald D.
King, Frank W.
Kyte, George W.
Law, John Alexander
Lunder, Daryl G.
McCaffrey, John
McCarthy, F. Dennis
McLennan, Robert E.
McLeod, John William D.

Mackin, William R.
Manz, Ronald L. L.
Mitchell, David William
Mullins, Euthan V.
Paicu, George
Panchysyn, Edward J.
Patterson, Frank W.
Pemberton, Richard L.
Peters, Gerard J.
Protopappas, Harry
Raczuk, Taras W.
Richards, Carl B.
Robinson, John L.
Rudd, Robert W.
Schulz, Johannes G.
Seabrook, Philip T.
Siemens, Donald L.
Smith, Tracy E.
Stein, Richard A.
Syms, Gordon H.
Szaszkiewicz, Joseph
Tetreau, Ernest M.
Thompson, Robert S.
Tym, Harvey Gerald
Webb, Kenneth C.
Webster, Stanley Leonard
Wright, Ralph D.

ALBERTA LAND SURVEYING

APRIL, 1961
- Armfelt, Richard
- Baker, Robert
- MacCrimmon, Morrison
- Rachansky, Bernard
- Rice, John D.
- Swanby, Thomas
- Tessari, E. J.

SEPTEMBER, 1961
- Fitzgerald, Leo
- Halahuric, Frank
- Price, Donald

LAW

APRIL, 1961
- Agrios, Jack Nicholas
- Anderson, Ernest Virgil
- Bradley, Norman A.
- Brownlee, John Stuart
- Bryan, John Alan
- Casey, John Charles
- Chomicki, Peter Roy
- Clarke, Ronald Victor
- Coady, Artur Francis
- Cummings, Ronald Gordon
- Decore, John Victor
- Denecky, Steve
- Filer, Sam Norman
- Gibbs, Reginald John
- Glenn, Alexander Irving
- Goss, Robert Fraser
- Horner, Frederick Garrick
- Horsman, James Deverell
- Hulme, George F.
- Hurov, Harvey Jerrold
- Jones, Frank Douglas
- King, Jack Avery
- Kuss, Daniel Reid
- Leefe, James Clifford
- Lerner, Stanley J.
- Lutz, Arthur Morton
- MacWilliams, Donald A.
- McIntyre, Paul C.
- McKillop, Duncan
- McKim, George F.
- Macklin, George Alexander
- Morrison, Glenn, M.
- Murray, Alec Thirlwell
- Patrick, Lynn Allen
- Pekarsky, Daniel Uri
- Pickard, Harry Edmison
- Piper, Frank Ivan
- Pittman, Gerald Wallace
- Ross, David James R.
- Sadownik, Rostyk
- Schwarz, Heinz
- Shymko, Nicholas
- Vickerson, Robert K.
- Wong, Rosalie Darlene
- Zaruby, Eugene Basil

AUGUST, 1961
- Rootes, Kennth James
- Szmolyan, Leslie Julius

NURSING

- Aakerberg, Ruth S.
- Achtemichuk, Elizabeth
- Adachi, Emiko
- Adamovich, Julia Pearl
- Adnitt, Darleen Mae
- Anderberg, Ellen M.
- Andersen, Grace E.
- Anderson, Ruby Doreen
- Arndt, Arlene Judith
- Arrowsmith, Sheila
- Ashdown, Viola Jean
- Audy, Solange, M. L.
- Austin, Darlene Fay
- Baerg, Marcella P.
- Baker, Leah Denise
- Bartlett, Linda Lou
- Bartlett, Nina Mae
- Bartling, Lucille O.
- Bateman, Sheila W.
- Bauer, Gloria Gladys
- Bauer, Eileen Teresa
- Baumbach, Arlene M.
- Baxter, Gloria Jean
- Bealing, Helen Sandra
- Behm, Jean Louise
- Bennett, Carol L.
- Bennett, Inez Marcia
- Bennett, Zella May
- Benoit, Joan Miriam
- Berkhold, Georgia I.
- Bertelsen, Ruth Christeen
- Bertrand, Janice L.
- Betts, Eleanor Mary
- Bichler, Mary Adelaide
- Bird, Sophie Helen
- Blackmore, Ada Rochelle
- Blackwell, Bette Joan
- Blackwell, June Anne
- Blain, Aline Beatrice
- Blakely, Joanne
- Block, Berna
- Bohnet, Viola M.
- Boisvert, Pauline B.
- Boness, Darla Ione
- Borden, Lynda Lee
- Bourbonnais, Rhea O.
- Bourgeois, Lorraine
- Bowers, Marg Pamela
- Boyce, Lorna J.
- Breault, Mariette Y.

Briand, Rose Marie
Briggs, Joan Laura
Brimacombe, Myrna M.
Brooke, Sheila Margaret
Brooks, Sandra M.
Browett, Marianna C.
Brown, Johann Hill M.
Bruce, Harriet Frances
Brunelle, Theresa M.
Brunt, Pamela Noreen
Bullen, Vivian May
Burch, Deloyc D.
Burchell, Rosemary Clare
Burns, Gavina Violet
Burr, Carolyn
Burris, Sheeran B.
Burton, Margaret Joan
Buttrey, Rita May E.
Cameron, Barbara L.
Cameron, Helen J.
Campbell, Margaret M.
Campbell, Shirley M.
Cardwell, M. Joanne
Carleton, Edna Gail
Carr, Joan Margaret
Caspar, Theresa E.
Castella, Carol Joyce
Caswell, Rosemarie
Chamberlain, Margaret A.
Chartrand, Denise Dena
Chayer, Darlene Isabel
Chen, Mae Clara
Cheng, Judith Q.
Church, Beverley Ann
Clark, Anne Patricia
Coles, Elizabeth A.
Coiijn, Hella G.
Common, Sandra Jean
Connell, Helen Louise
Cote, Juliette Silvia
Cotton, Doreen Freda
Court, Marlene Lavon
Cowley, Anne E.
Coxon, Lillian
Crabb, Jennifer P.
Crick, E. Darlene
Criddle, Anne D.
Cunnings, Constance
Currie, Barbara Joan
Curry, Myrna Eileen
Cyre, Marcelle Monique
Dach, Elizabeth E.
Dangerfield Verna F.
Davidson, C. Dianne
Day, Ansley E.
Deputan, Sylvia Nancy
Dick, Katherine W.
Docking, Donna Mae
Dombroski, Florence
Donaghue, Shirley
Donaldson, Jean M.
Doroshenko, Leslie A.
Drabyk, Darlene M.
Draper, Margaret H.
Dulsrud, Darlene M.
Dunn, Beatrice Jean

Eberhardt, Geraldine
Eccleston, Sharron P.
Edgar, Gloria J.
Eirich, Helen Dorothy
Elliot, Maryanne M.
Ellmers, Jacqueline
English, Gladys E.
Evans, Beverly M.
Fairley, Andrea Laverne
Fedorak, Victoria
Felske, Veronica L.
Fengstad, Elvina Ann
Ferguson, Sheila K.
Ferrell, Judith Kay
Finch, Joan Eleanor
Fisher, Judith Lucille
Fjeldheim, Gloria M.
Flynn, Barbara Ann
Footz, Loretta June
Forsyth, Dixie Jean
Foster, Barbara Mae
Francey, Marjory Jean
Frizzell, Alice M.
Fullerton, Velma L.
Furuse, Mutsuko
Fyk, Violet Madjah
Gaetz, Kathryn E. J.
Galloway, Joan Mary
Gammon, Barbara May
Gannon, Gloria, R.
Gant, Marna Denise
Gau, Marie Theresa
Gawn, Maureen Currie
Gaychuk, Eleanor Margaret
Gerrish, Sally
Gibson, Marilyn D.
Gibson, Ruth Eleanor
Gilmour, Florence A.
Godsell, Marjorie May
Goodbrand, Ruth J.
Goulet, Joyce Marilyn
Graff, Doreen Deanna
Greidanus, Theresa
Greig, Elnora Anne
Groenendyk, Anne
Hagel, Marie Theresa
Hagenson, Loretta Ada
Halland, Arlene E.
Hamabata, Nadine
Hammel, Magdalena F.
Hanna, Phylis Elaine
Hanson, Phyllis, M.
Hartry, Ellen Lorraine
Hartry, Violet Elaine
Haslam, Donna Elaine
Haslam, Myrna Darlene
Hawkins, Margaret P.
Hay, Elizabeth M.
Haydamack, Jean A.
Hayter, Rita Diane
Hazlett, Sally W.
Heather, Marcille F.
Henderson, Beverlee
Heney, Sonja Ellen
Hering, Olga
Hertzog, Lorraine V. F.

Hickey, Mary
Higgs, Marilyn
Hildebrandt, Hilde
Hilgartner, Theresa O.
Hilliard, H. Irene
Hines, Sheila Diane
Hirsch, Katherine M.
Hirsche, Marjorie K.
Hodson, Jean
Holubisky, Sylvia Z.
Holwegner, Mona Ruth
Horricks, Deanna J.
Houston, Lorraine B.
Hughes, Dorothy N.
Hughes, Lorna Beth
Husler, Joan
Hydorn, Linda Gail
Inkster, Beverly B.
Ireland, Audrey M.
Ireland, Glenda J.
Ivins, Patricia J.
Jacobsen, Kirsten M.
Jaster, Marjorie J.
Jensen, Carole J.
Jickling, Laurene I.
Johnson, Victoria V.
Johnston, Norma Diane
Jones, Maureen Ann
Jorgenson, Betty L.
Juba, Rose Marie
Kaiser, Elizabeth M.
Kamp, Kirsten Emilie
Kanashiro, Mihoko S.
Kelly, Kathleen Joan
Kendze, Deanna M.
Kennedy, Mary J.
Kennedy, E. Beverley
Kennedy, Margaret R.
Kerber, Friederike H.
Kerr, Carol Ann
Kluzak, Helen M.
Knowles, Ruth Patricia
Kocher, Marion G.
Kole, Lorna G.
Koroluk, Nadie
Kozack, Lillian
Kozak, Katherine D.
Kozaniuk, Florence A.
Kroeker, Beverly Ann
Kroeker, Katherine E.
Kroetch, Blanche T.
Kruger, Lily Marie
Krumpic, Joyce D.
Kunigiskis, Adele E.
Kuriantnyk, Josepha S.
Kurylowich, Bertha
Langlois, Margaret A.
Lasecki, Margaret S.
Lavalley, Vivienne G.
Leitch, Sandra C.
Lemon, Margaret D.
Leonard, Daryl A.
Lepper, Marlene M.
Lesnik, Helen D.
Levy, Roslyn E.
Liese, Hildegard

Lorens, Adele K.
Loucks, Alma Noreen
Luchka, Olga
Lundberg, Georgie C.
Lyon, Eileen M.
MacDonald, Mary B.
MacKay, Beatrice Ann
MacKenzie, Barbara Ann
MacLeod, Faith D.
MacLeod, Valerie May
MacMillen, Ann Margaret
MacPherson, Maxine A.
MacPherson, Sheila Gail
MacQuarrie, Helen C.
Maguire, Sharon Rose
Maher, Patricia Ann
Maksymiw, Lucia
Marshall, Violet A.
Matthews, Marjory R.
Mattson, Juane Elsie
Matty, Dianne Louise
May, Shelia Marion
McAdam, Carol Mae
McBrien, Patrica Anne
McCauley, K. Audrey
McCourt, Ruth L.
McCullough, Lillian F.
McGrath, Isabelle M.
McGinnis Norma Ruth
McIlroy, Alice E.
McKerihan, Jean M.
McKerihan, Linda Ann
McKerricher, Elaine
McKibbon, Stella M.
McKiel, Ruby Elaine
McLean, Gloria E.
McLeay, Jean E.
McLeod, Marlene C.
McManus, Noreen M.
McNicoll, Patricia A.
Meclaw, Sylvia M.
Meier, Bernice S.
Meighen, Elaine H.
Melsness, Audrey G.
Meyer, Robert
Michlich, Marianne J.
Mickla, Jean E.
Millbank, Georgia
Morgan, June L.
Morin, Andrea L.
Morin, Florence M.
Mraz, Teresa R.
Muraca, Arlene H.
Murhpy, Lorna M.
Mutch, Ferne E.
Neely, Janet Helen
Nelner, Joyce Ina
Newnham, Eleanor J.
Newton, Marla N.
Nicholson, E. Anne
Niedswiecki, Hazel L.
Nielsen, Anne Marie
Nielsen, Olive L.
Nimchuk, Cecelia G.
Nimchuk, Patrica V.
Niwa, Marjorie A.

Nolan, Sherry C.
Nyback Sonja D.
Nysetvold, Madeleine A
Ogden, June A.
Ogle, Mary Louise
Olson, Delores E.
O'Neil, Beverly June
O'Neill, Sally W.
Osinchuk, Juan A.
Ostberg, Marilyn G.
Palas, Vera
Palas, Vilma
Park, Janet Diane
Parsonage, Bertha L.
Paterson, Sharon F.
Patrick, Hazel W.
Patteson, Lynda D.
Pchelnyk, Lucy
Pedersen, Ann Marie
Pedersen, Eileen M.
Perchinsky, Deanne M.
Perka, Margaret B.
Perka, Rosemarie P.
Perrin, Marjorie A.
Peszat, Maurren J.
Peters, Ruth Kennedy
Petersen, Carol Anne
Phillips, Donna L.
Piebiak, Joan A.
Piggott, Dorothy Marie
Platzer, Marilyn
Plouffe, Blanche E.
Pogson, Geraldynne M.
Poland, Florence M.
Popyk, Barbara Anne
Porter, Audrey Fay
Price, Janet Ann
Price, Sandra J.
Provencher, Denise M.
Pytel, Doreen Alice
Quon, Phyllis Mabel
Radomsky, Evelyn E.
Reti, Alice M.
Richard, Diane B.
Ritchie, Kay Arlene
Rivet, Antonia R.
Roberts, Florence
Robertson, Sharon Meta
Robinson, Carol J.
Rochow, Reinhilde M.
Rodbourne, Diane M.
Rodwell, Frances L.
Royer, June E.
Ruks, Edna
Rumen, Ksenia V.
Ruschienskey, Jane E.
Rutschke, Melva J.
Sakowski, Marilyn A.
Saliwonchuk, Olive A.
Savill, Belinda
Scheidl, Shirley A.
Schick, Helen M.
Schlachter, Lenora H.
Schnell, Marilyn E.
Schon, Marie E.
Schuller, Teresa M.

Scotland, Dorothy M.
Scott, Moreen
Semeniuk, Katherine A.
Seymour, Annette
Shannon, Elizabeth A.
Shannon, Perry Lynne
Shantz, Shirley J.
Shaw, Eleanor Judith
Sheppard, Judith A.
Sherwin, Sylvia M.
Shipley, G. Jolene
Short, Darilyn Jean
Shyry, Zonia S.
Sills, Darlene L.
Sitler, Darlene L.
Skoreyko, Sylvia J.
Skrove, Mary Louise
Slavik, Mary B.
Sloat, Mary Eileen
Smalley, Marjorie A.
Smart, Carolyn M.
Smith, Esther A.
Smith, Margaret M.
Snider, Ruth C.
Soice, Marjorie M.
Sokvitne, Shirley M.
Sokvitne, Signe M.
Spornitz, Paula F.
Stark, Patricia Ellen
Stawart, Margaret Anne
Steele, Marjorie Ann
Stevens, Margaret L.
Stewart, Patrica J.
Stewart, Edith Marilyn
St. Martin, Margaret Rose
Stuber, Mable Marion
Stuckey, Patricia D.
Sweet, Marilyn J.
Teulon, Margaret M.
Theissen, Elsie J.
Thompson, Sherill G.
Thompson, Virginia R.
Thompson, Winnifred
Thornton, Elizabeth B.
Thumlert, Caroline M.
Tiedemann, Anita J.
Tiessen, Verna
Tilley, Bonnie B.
Tobler, Jeanette R.
Todd, Marilyn J.
Tollestrup, Barbara
Train, Isabel Prentice
Tudor, Donna O.
Turbak, Rosalie P.
Turcato, Elda J.
Ulmer, Gertrude Louise
Upshall, Ruth M.
Vanderaegen, Jeannette
Verenka, Marie Anne
Vincent, Gwendolyn B.
Wacko, Stephanie M.
Waddington, Joan R.
Walker, Corrine, A.
Wallace, Leola M.
Warlo, June M.
Warwick, M. Gail

Weatherhead Treva C.
Weis, Norma Olive
Wells, Alma Frances
Weme, Caroline M.
Wesley, Alice Joy
Whitby, Catherine Jean
White, Lucinda M.
Wiens, Helen
Wiley, Mildred Edith
Wilkie, Virginia H.

MAY, 1961
Enderud, Doris

JUNE, 1961
Boylan, Daryl Elaine

JULY 1961
Kooy, Peter Melvin

SEPTEMBER, 1961
Hunka, Evelyn Blanche

APRIL, 1961
Kozorowski, Emil W.

AUGUST, 1961
Beck, Byron Edward
Roberts, Henry Gerald

SEPTEMBER 1961
Adams, Francis Loyal

Wilkinson, Carole A.
Williams, Loretta
Wilson, Irene Isabella
Wilson, Margaret M.
Wolfe, Norma R.
Wotherspoon, Elizabeth H.
Wright, Melba Louise
Wright, Theresa A.
Young, Margaret Eva
Ziebart, E. Barbara

PHARMACY

Enderud, Harold K.

VETERINARY MEDICINE

Stephens, Harold Allen

Hill, Everrett Doyle

Uncl Mo. Mexx. have not invite Rita & Maggie Wilson who was coming?

CPSIA information can be obtained
at www.ICGtesting.com
Printed in the USA
BVHW041525221121
622229BV00009B/266